MARVELOUS MULTIPLICATION
AND DAZZLING DIVISION

Author
Judith Hillen

Editor
Betty Cordel

Illustrator
Reneé Mason

Desktop Publisher
Tracey Lieder

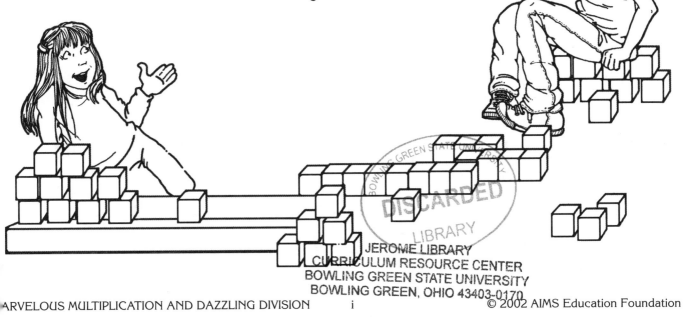

MARVELOUS MULTIPLICATION AND DAZZLING DIVISION

This book contains materials developed by the AIMS Education Foundation. **AIMS** (**A**ctivities **I**ntegrating **M**athematics and **S**cience) began in 1981 with a grant from the National Science Foundation. The non-profit AIMS Education Foundation publishes hands-on instructional materials (books and the quarterly magazine) that integrate curricular disciplines such as mathematics, science, language arts, and social studies. The Foundation sponsors a national program of professional development through which educators may gain both an understanding of the AIMS philosophy and expertise in teaching by integrated, hands-on methods.

Copyright © 2002 by the AIMS Education Foundation

ISBN 978-1-932093-01-8

Printed in the United States of America

DEDICATION AND AUTHOR'S NOTE

The experiences here are intended to be of encouragement and support for teachers and young learners everywhere as they seek to understand and enjoy basic operations on whole numbers.

This publication is dedicated to eight special children and others just like them in hopes that they will love learning about math.

To: Kyle, Emily, Ryan, Grant, Ellie, Spencer, Katie and Jon

With love,
Judith A. Hillen

A SPECIAL THANKS

Out thanks to Dr. Arthur Wiebe and Wilbert Reimer for their ideas and expertise for the development of many of the included experiences in this publication.

Marvelous Multiplication and Dazzling Division

Table of Contents

I Hear and I Forget,

I See and I Remember,

I Do and
   I Understand.

Chinese Proverb

INTRODUCTION AND OVERVIEW

This collection of hands-on experiences seeks to address both the conceptual understanding of the processes of multiplication and division as well as the procedural proficiency of calculating multi-digit operations.

Five major ideas define the focus of this publication:
- Building conceptual understanding
- Playful, intelligent practice
- Historical connections
- Problem solving applications
- Multiple assessment strategies

NCTM STANDARDS 2000*

Number and Operations
- *Understand the place value structure of the base-ten number system and be able to represent and compare whole numbers and decimals*
- *Recognize equivalent representations for the same number and generate them by decomposing and composing numbers*
- *Understand various meanings of multiplication and division*
- *Understand the effects of multiplying and dividing whole numbers*
- *Identify and use relationships between operations, such as division as the inverse of multiplication, to solve problems*
- *Develop fluency with basic number combinations for multiplication and division and use these combinations to mentally compute related problems, such as 30 x 50*
- *Develop fluency in adding, subtracting, multiplying, and dividing whole numbers*
- *Develop and use strategies to estimate the results of whole number computations and to judge the reasonableness of such results*
- *Select appropriate methods and tools for computing with whole numbers from among mental computation, estimation, calculators, and paper and pencil according to the context and nature of the computation and use the selected method or tool*

Algebra Standard
- *Describe, extend, and make generalizations about geometric and numeric patterns*
- *Represent and analyze patterns and functions, using words, tables, and graphs*
- *Model problem situations with objects and use representations such as graphs, tables, and equations to draw conclusions*

Communication Standard
- *Recognize and use connections among mathematical ideas*
- *Recognize how mathematical ideas interconnect and build on one another to produce a coherent whole*

Representation Standard
- *Create and use representations to organize, record, and communicate mathematical ideas*
- *Use representations to model and interpret physical, social, and mathematical phenomena*
- *Select, apply, and translate among mathematical representations to solve problems*

Problem Solving Standard
- *Build mathematical knowledge through problem solving*
- *Solve problems that arise in mathematics and in other contexts*
- *Apply and adapt a variety of appropriate strategies to solve problems*
- *Monitor and reflect on the process of mathematical problem solving*

* Reprinted with permission from *Principles and Standards for School Mathematics*, 2000 by the National Council of Teachers of Mathematics. All rights reserved.

MAKING SENSE OF WHOLE NUMBER OPERATIONS

LEVELS OF OPERATIONS

Concrete/Manipulative
Sets of objects are combined or separated and counted or measured in order to experience the basic operations in a concrete way.

Representational/Pictorial
Objects and actions are pictured and counted or measured.

Abstract/Symbolic
Numerals, symbols, and relationship signs are used to represent objects and actions at the concrete and representational levels.

MAKING SENSE OF WHOLE NUMBER OPERATIONS

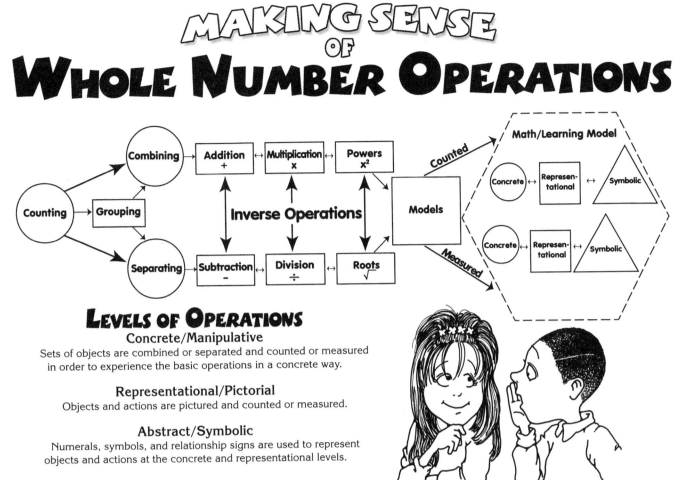

LEVELS OF OPERATIONS

Concrete/Manipulative
Sets of objects are combined or separated and counted or measured in order to experience the basic operations in a concrete way.

Representational/Pictorial
Objects and actions are pictured and counted or measured.

Abstract/Symbolic
Numerals, symbols, and relationship signs are used to represent objects and actions at the concrete and representational levels.

Introduction

Traditionally the teaching of number concepts *begins* with introducing students to arithmetic through numerals and symbols. What elementary students need most is to make *meaning* of number and the basic operations through the use of concrete manipulatives and then make connections between the mathematics they study at the concrete level and the numerals and symbols at the abstract level. The difficulties that many children have with mathematics, specifically the potential for computational errors, are due in large part to their inability to make sense of the numeric symbols and the connection of any meaning to the algorithm.

The accompanying diagram speaks to us about a view of arithmetic that deals with groups. Students enter school and begin to learn to count by recognizing a one-to-one correspondence between an object and a number. Soon they recognize the number of objects in small groups without having to count each one. It is this recognition of groups that paves the way for considering each of the operations as one of combining or of separating groups of objects. Addition, multiplication, and raising to a power are all examples of a process of combining or joining while subtraction, division, and extracting a root are examples of a process of separating or partitioning. Each of the basic operations should be experienced at a variety of levels.

Beyond Understanding — The Basic Facts

A balanced mathematics program includes frequent doses of *playful, intelligent practice,* and *creative, real-world problem solving* experiences that provide opportunity to apply the basic operations.

Building Conceptual Understanding
Multiplication

Concrete/Manipulative Level

At this level students join equal sets of objects in order to experience the basic operation in a concrete way.

Using countable objects

With countable objects, such as buttons or bows, the multiplication operation consists of joining two or more *equal* sets of objects to form one set. At the manipulative stage students construct equal sets of objects and then join them into one larger set. The move to combine *equal* sets into one is referred to as repeated addition or multiplication. By counting, students can determine the number of objects in each set and in the combined set—confirming that no objects have been lost or added in the transaction.

An array model

Using countable objects, such as bottle caps or coins, students may arrange equal sets of these in rows and columns to form a rectangular array. The perpendicular arrangement of these equal sets invites students to consider the number of objects in *one* row and the number of rows to determine the total number by repeated addition or multiplication.

Using measured objects

Objects such as blocks or small boxes may be used for a measured or area model for multiplication. To multiply these sets, we place the objects in rows and columns to complete a rectangle. To determine the area of the rectangle, we multiply the length of a column of blocks times the length of a row of blocks.

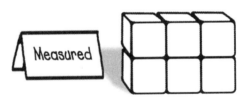

Representational/Pictorial Level

At the representative level, sometimes called the connecting stage, objects and actions are represented or depicted by pictures or diagrams. For multiplication equal sets are pictured and arrows indicate that they are combined.

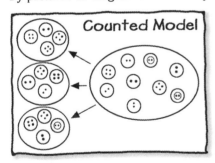

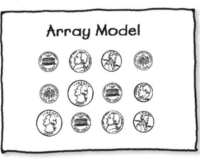

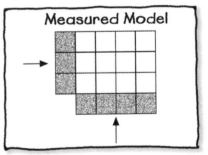

Abstract/Symbolic Level

At this level numerals, symbols, and relationship signs are used to represent the objects and actions of the concrete and representative levels. At the symbolic stage, the number of objects or measured units in *one* set are counted and recorded. The number of sets is counted and recorded and joining them all together is represented by the symbol x. The = sign indicates that no objects were added or lost in the transaction.

4 sets of 5 objects = 20 objects

4 x 5 = 20

5 sets of 4 objects = 20 objects

5 x 4 = 20

Sets by the Number

Topic
Warm-up experience to introduce the idea that objects come in sets

Learning Goals
Students will:
- build capacity for understanding the meaning of multiplication as the union of equal sets; and
- observe, search, discover, and imagine objects or events that come in sets.

Materials
Sets by the Number—A Scavenger Hunt activity page
10 index cards (5" x 8") per team

Management
1. To heighten motivation or excitement, place a time limit and suggest that it is a contest.
2. Place students in teams of two or three. Ask each team to find as many sets as they can in each category.

Procedure
1. Distribute scavenger search *Sets by the Number* to students.
2. Direct students to name, describe or picture the "set" on a 5-inch by 8-inch index card.
3. Build a class bulletin board of Sets by the Number. Display cards to show how frequently sets of the same number occur.

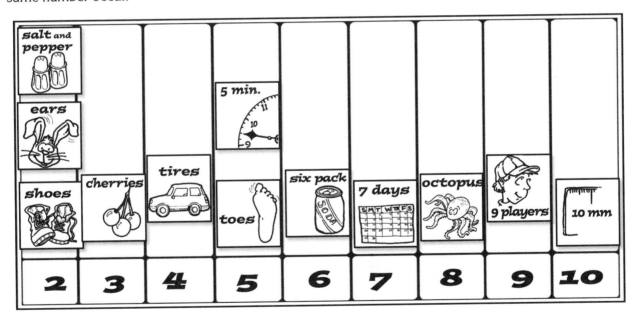

Discussion
1. Why do you think objects are packaged or bundled in sets? [for convenience, for sales, for efficiency]
2. What are the smallest sets you were able to find? …the largest? (Answers will vary.)
3. Describe unusual sets you found. [Example: Life Savers come in sets of 14.]
4. Sometimes events come in sets too. What events can you think of that come in sets? [Sporting events: nine innings to a baseball game; four quarters for a football or basketball game.]

Sets by the Number
-A Scavenger Hunt

Name, describe, and picture each set you find on one index card. Your team will need 10 cards.

1. Find three things that come in pairs.

2. Name one thing that comes in threes.

3. Find two things outdoors that come in fives.

4. Find something in the classroom that comes in sets. How many in each set?

5. Think about something at home or at the store that is bought in sets of six.

6. What comes in nines?

7. Picture the most unusual set you found.

Sets by the Number
-A Scavenger Hunt

Name, describe, and picture each set you find on one index card. Your team will need 10 cards.

1. Find three things that come in pairs.

2. Name one thing that comes in threes.

3. Find two things outdoors that come in fives.

4. Find something in the classroom that comes in sets. How many in each set?

5. Think about something at home or at the store that is bought in sets of six.

6. What comes in nines?

7. Picture the most unusual set you found.

Stacking the Facts

Topic
Whole number operations
Area model for multiplication
Multiplication facts

Learning Goals
Students will:
- develop fluency with single digit multiplication facts, and
- use area model to provide meaning for the multiplication operation

Guiding Document
NCTM Standard 2000*
- *Describe, extend, and make generalizations about geometric and numeric patterns*

Materials
Per small group:
6 2-cm grid sheets (see *Management 1* and *2*)
scissors
25 plain colored index cards (5" x 8")
plastic bag (see *Management 3*)

For each student:
1 2-cm grid
1 *Multiplication Table* grid

For the class:
stapler

Background Information
In this activity students will recognize the multiplication table as an area model by first making square "table tops" and recording their dimensions of length and width and total number of squares. They will then construct oblong or rectangular "table tops" and record their dimensions and total to complete the multiplication table of facts up to 9 x 9.

This activity has potential for connections of at least two kinds: literacy connections and historical connections.

Literacy Connections: There are several good trade books that are worth sharing with students in middle grades. One is *Amanda Bean's Amazing Dream* by Marilyn Burns (Scholastic Press: New York, 1998.) In this delightful story, Amanda Bean, who likes to count everything, discovers that being able to multiply will help her count everything faster. Another book that is

especially helpful for practice counting and squaring numbers from one to ten, is *Sea Squares* by Joy Hulme (Hyperion Paperbacks for Children, 1991). This book includes rhyming text and illustrations of sea animals.

Historical Connections: Jakow Trachtenberg, a Russian-born Jew, made a significant contribution to mathematical education through the design of a "speed system" of basic computation. See *Square Rules*.

Management
1. Copy the 2-cm grid paper onto plain paper in one color and distribute one page to each student for *Part One* of the activity, *Building Arrays*.
2. Copy the 2-cm grid paper onto six different colors of paper. Have extra copies available for students that make errors in cutting.
3. A collection of nine different squares needs to be cut from the colored grid paper. Squares measure 1 x 1, 2 x 2, 3 x 3, 4 x 4, 5 x 5, 6 x 6, 7 x 7, 8 x 8, and 9 x 9. These squares may be cut ahead of time by the teacher or prepared by the students as part of the activity. Oblong or rectangular pieces will need to be cut from the pieces of grid paper remaining after the squares are cut.
4. For ease of distribution, package the grid paper in plastic bags.
5. Copy one 2-cm grid paper labeled *Multiplication Table* to each student.

Procedure
Part One —Building Arrays
1. Distribute one copy of the 2-cm grid paper to each student. Introduce the idea of building and cutting out rectangular arrays as a warm up experience that connects to the idea of multiplication as an area model. Have students cut out a 2 by 3, a 2 by 5 and a 2 by 7 array.

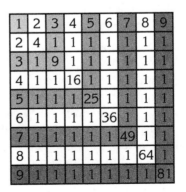

2. Continue with some 3 by and 4 by arrays. Discuss the idea that these arrays represent an area covered with square units.

3. Explain that the array can be described as a multiplication fact that counts the number of squares in a row or column multiplied by the number of rows or columns respectively. For example, a 2 by 3 represents an area that is 2 units wide and 3 units long and covers an area of 6 units. This can be written as 2 x 3 = 6.

Part Two—Stacking the Facts

1. Distribute one bag of nine colored square grid papers to each small group of students.

2. Have the students scatter the paper squares on the table.

3. Tell them to imagine that these are square tables, like arrays, of different sizes and that they are going to be stacked gradually with the largest on the bottom and the smallest on top. Explain that no table may be stacked on another until each corner is labeled with a number. The upper left corner of every table is labeled 1 and provides the common point at which all tables are attached. The opposite corner, bottom right, indicates the total number of square tiles on the table top. The remaining two corners indicate how many tiles in the row or column and are labeled accordingly. For example, one table may be as pictured.

1		3
3		9

4. Guide the students in sequencing the completed tables face-up, largest on the bottom to smallest on top, and staple the set at the number one corner.

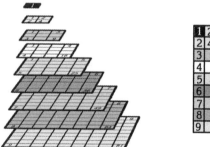

5. Tell the students to transfer the numbers from the square table data to the empty 9 x 9 grid provided.

6. Now direct them to cut oblongs or rectangles from the remaining colored grid paper.

7. Have them label the corners of the new table tops in the same way as the square tables were done.

8. Tell them to **stack** the new oblongs on the numbered table and fill in remaining empty corners.

9. Have the students transfer the data to the grid.

10. Ask students to choose five multiplication facts from the table that are the most difficult to remember quickly. Have them cut a blank index card along the diagonal to produce triangular flash cards. Tell them to use a black marker to label each corner of the card with part of the number fact. The largest number goes in the right square corner and the factors in the remaining corners. Have them place appropriate operations signs between each pair of numbers. For example, 7 x 8 = 56 would appear as shown.

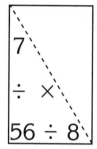

11. Encourage students to use the cards to practice those facts with a buddy. Tell students to cover any number and give the answer to the corresponding number fact. Example: Cover 56; What is 7 x 8? Cover 8; What is 56 ÷ 7?

Discussion

1. Ask students to describe the square papers in terms of their similarities and differences. [different sizes, different colors; same shape; all sides on each square same length.]

2. What patterns did you observe in the square tables? [Square numbers run diagonally from upper left to lower right corner.]

3. What shapes make up the tables that are not square? [oblongs or rectangles]

4. How are the oblongs or rectangles the same or different from the square tables? [One dimension in the oblongs is longer or shorter than the other dimension. They are all area models for multiplication.]

5. What is the relationship between the numbers in the four corners? [one table; opposite corner is the product; other two corners are the factors that represent the length and width of the area model]

6. What patterns can you find that would be helpful in remembering the multiplication facts more easily?

Evaluation

1. Take Ten—Quick Quiz of ten facts with a partner. Choose any ten facts and quiz your partner. Encourage students to select difficult facts. Partners score each other.

Evidence of Learning

1. Improvement of scores in quizzes over time.

* Reprinted with permission from *Principles and Standards for School Mathematics,* 2000 by the National Council of Teachers of Mathematics. All rights reserved.

Stacking the Facts

Multiplication Table

Topic
Multiplication, Area

Learning Goals
Students will:
- recognize the role of a base-ten number system in multiplication,
- . be able to visualize the process of multiplication as an area model of covering a rectangular region, and
- visualize a pattern of partial products connected to place value concepts.

Guiding Document
NCTM Standards 2000*
- *Understand various meanings of multiplication and division*
- *Understand the effects multiplying and dividing whole numbers*
- *Understand and use the properties of operations such as the distributivity of multiplication over addition*

Materials
Base Ten Blocks
Colored pencils
$\frac{1}{2}$ cm grid paper
Rulers or straight edges
Student sheets

Background
In this experience, students will model the multiplication operation by covering a rectangular array with Base Ten Blocks. In this area model, two factors define the perpendicular dimensions of the rectangle, and the product is represented by the total number of cubes that cover the area. When the ones (units), tens (longs), and hundreds (flats) are used to construct the rectangular region, counting the area is easily accomplished. Furthermore, students are able to make sense of the partial products as they are represented visually.

Management
Under ideal conditions each student should have a set of Base Ten Blocks. If materials are limited, a small group of two to four students can share a set of blocks.

Procedure
1. Give students sample rectangles (*Facts First*) that represent some multiplication facts.
2. Discuss features that are common to all the samples, eliciting the ideas that all are rectangles, all have two dimensions (length and width), and the product of each is simply the total number of small cubes it takes to cover the area.
3. Ask students to cover the rectangles with the fewest possible Base Ten Blocks. Discuss how they determined the area of the rectangle. [They may share that they skip counted by some factor of the total.]
4. Distribute *Area Codes* and provide guidelines for filling the frames.
   a. Fill from the lower left corner where an X appears in a circle.
   b. Use the fewest number of pieces by filling first with flats, then longs, then units.
5. Ask students to cover each rectangle and to trace the flats, longs, and units in their position as a record of covering. Encourage them to look for patterns of arrangements and to find a way to quickly count the squares. Ask them to consider different counting methods for determining the area.
6. Direct them to include in their rectangles the following:
   a. the dimensions of the rectangle in expanded form [10 + 7 x 10 + 4]
   b. the dimensions and the product for each interior region

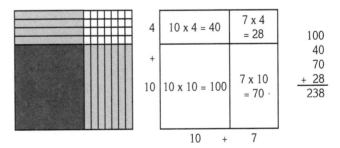

7. Complete *Area Codes*.
8. Extend the thinking by suggesting that students work on a problem by designing the rectangle to fit the problem. Within each small group, direct one student to suggest a two-digit multiplication problem for the group to build and solve with Base Ten Blocks. Have them record the problems

on large chart paper. Be sure that each team member designs a problem and is responsible for determining if the solution is correct.

9. Try using the inverse procedure to build a division problem. Give students the total number of cubes (the dividend) and either the divisor or the quotient. Have them build the rectangle to determine the missing dimension. For example, try 156 divided by 12.

Discussion

1. What do rectangles and multiplication have in common? [Both represent the combination of equal sets, and the two dimensions of the rectangle also represent the two factors that determine a product, or area, of the rectangle.]

2. What are multiplication facts and why is it useful to know them? [Multiplication facts generally represent totals of multiples of equal sets. They are useful as a tool for more rapid calculation of larger numbers.]

3. What strategies did you use to count up the number of cubes rapidly? [flats by 100, longs by 10, and units by 1]

4. What patterns did you find in how the Base Ten Blocks were placed in each rectangle? [Flats, the largest pieces, are always anchored in the lower left corner. A rectangle of longs could be placed horizontally in the upper left corner and also vertically in lower right corner. Units appear in the upper right corner.]

Evaluation

Distribute *Area Maps* to each group of students. Using half-centimeter grid paper, students are to represent in four colors, a rectangle for each problem given. Each interior region must be labeled and the dimensions identified.

Solutions

Following are the solutions for *Facts First*.

1. 4 x 3 = 12
2. 2 x 9 = 18
3. 8 x 3 = 24
4. 7 x 6 = 42
5. 4 x 5 = 20

Following are the solutions to *Area Codes*.

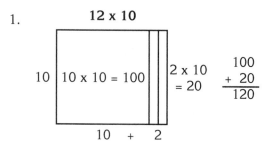

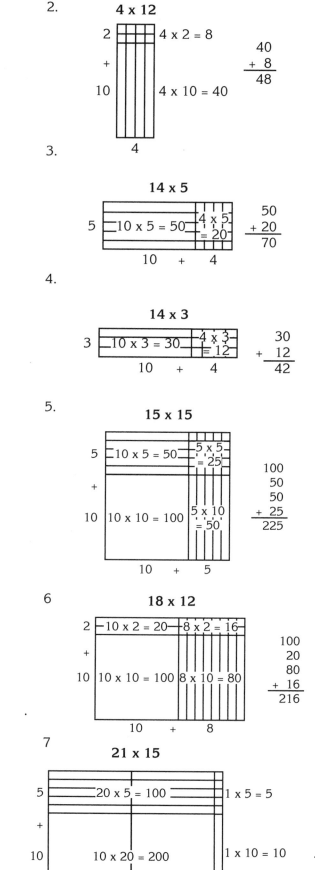

8.   23 x 14

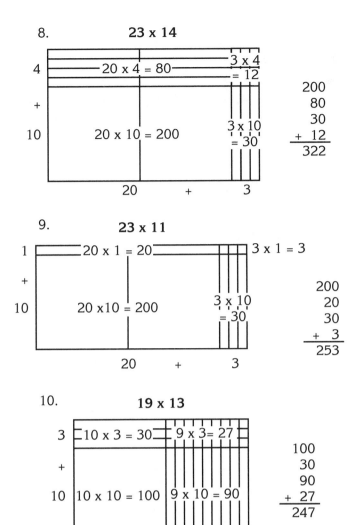

4 | — 20 x 4 = 80 — | 3 x 4 = 12
+ |
10 | 20 x 10 = 200 | 3 x 10 = 30
20 + 3

200
80
30
+ 12
322

9.   23 x 11

1 | 20 x 1 = 20 | 3 x 1 = 3
+ |
10 | 20 x 10 = 200 | 3 x 10 = 30
20 + 3

200
20
30
+ 3
253

10.   19 x 13

3 | 10 x 3 = 30 | 9 x 3 = 27
+ |
10 | 10 x 10 = 100 | 9 x 10 = 90
10 + 9

100
30
90
+ 27
247

Following are the solutions to *Area Maps*.

1.

25 x 13

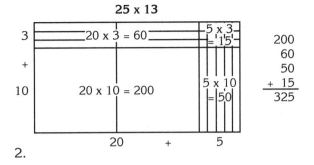

3 | 20 x 3 = 60 | 5 x 3 = 15
+ |
10 | 20 x 10 = 200 | 5 x 10 = 50
20 + 5

200
60
50
+ 15
325

2.

12 x 14

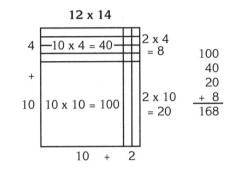

4 | 10 x 4 = 40 | 2 x 4 = 8
+ |
10 | 10 x 10 = 100 | 2 x 10 = 20
10 + 2

100
40
20
+ 8
168

3.

7 x 17

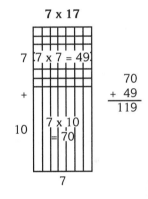

7 | 7 x 7 = 49
+ |
10 | 7 x 10 = 70
7

70
+ 49
119

4.

23 x 28

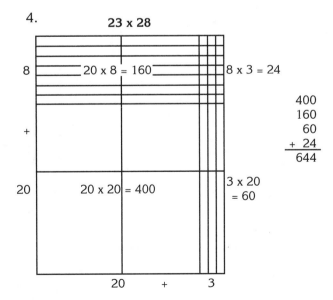

8 | 20 x 8 = 160 | 8 x 3 = 24
+ |
20 | 20 x 20 = 400 | 3 x 20 = 60
20 + 3

400
160
60
+ 24
644

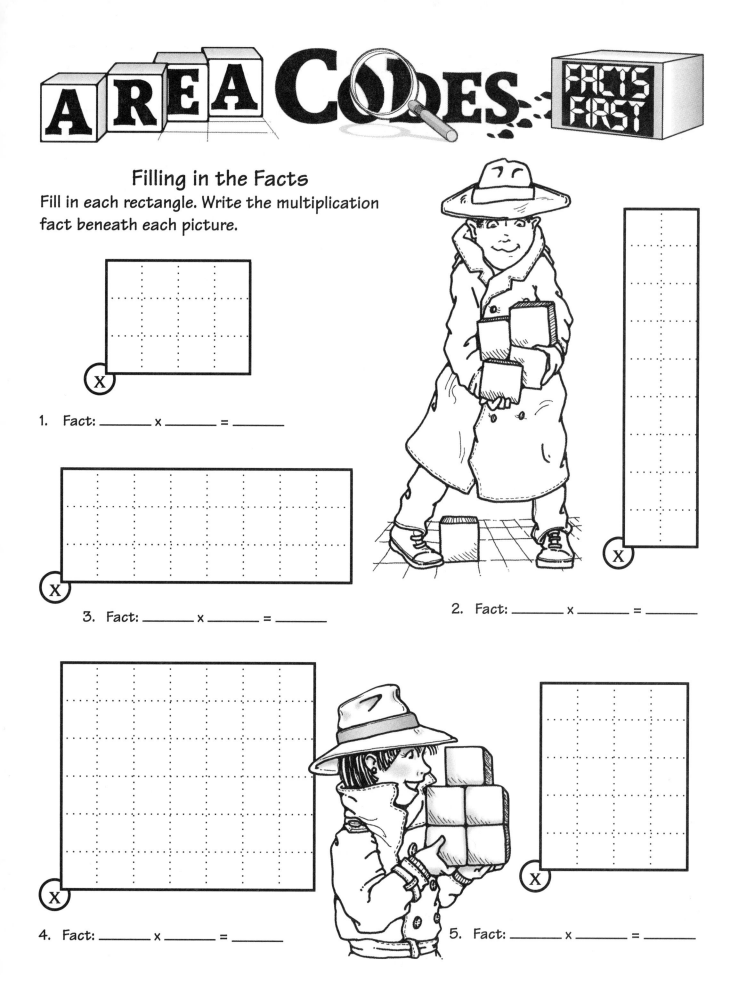

Filling in the Facts

Fill in each rectangle. Write the multiplication fact beneath each picture.

1. Fact: _____ x _____ = _____

2. Fact: _____ x _____ = _____

3. Fact: _____ x _____ = _____

4. Fact: _____ x _____ = _____

5. Fact: _____ x _____ = _____

AREA CODES

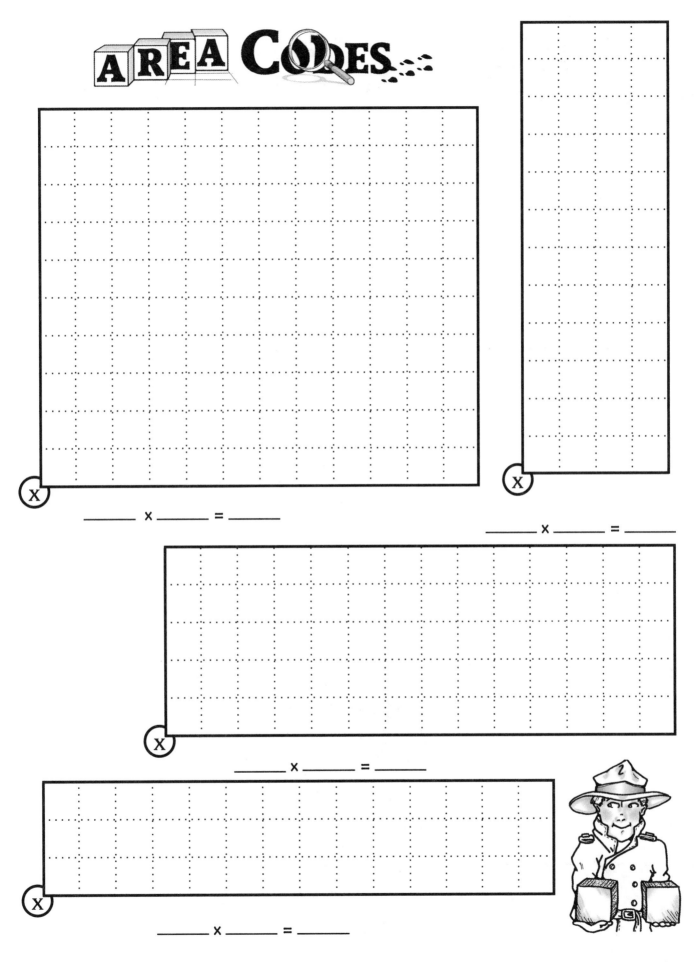

_____ x _____ = _____

_____ x _____ = _____

_____ x _____ = _____

_____ x _____ = _____

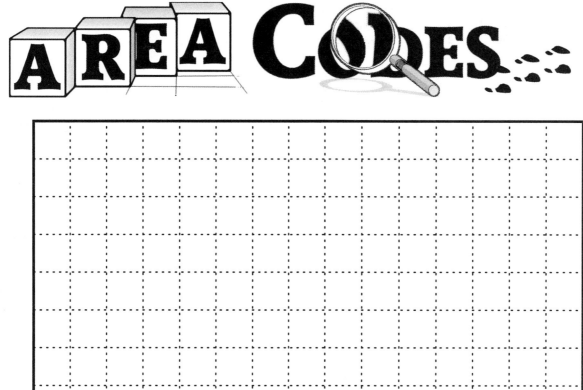

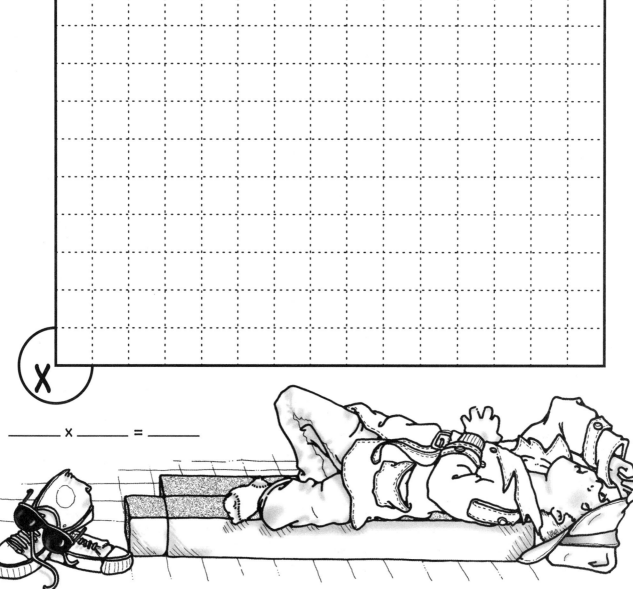

_____ x _____ = _____

X

___ × ___ = ___

AREA MAPS

Draw area maps for these problems using half-centimeter grid paper. Shade and label each region. Record the dimensions outside each area map.

25 x 13   7 x 17   12 x 14   23 x 28

AREA MAPS

Shade and label each region. Record the dimensions outside each area map.

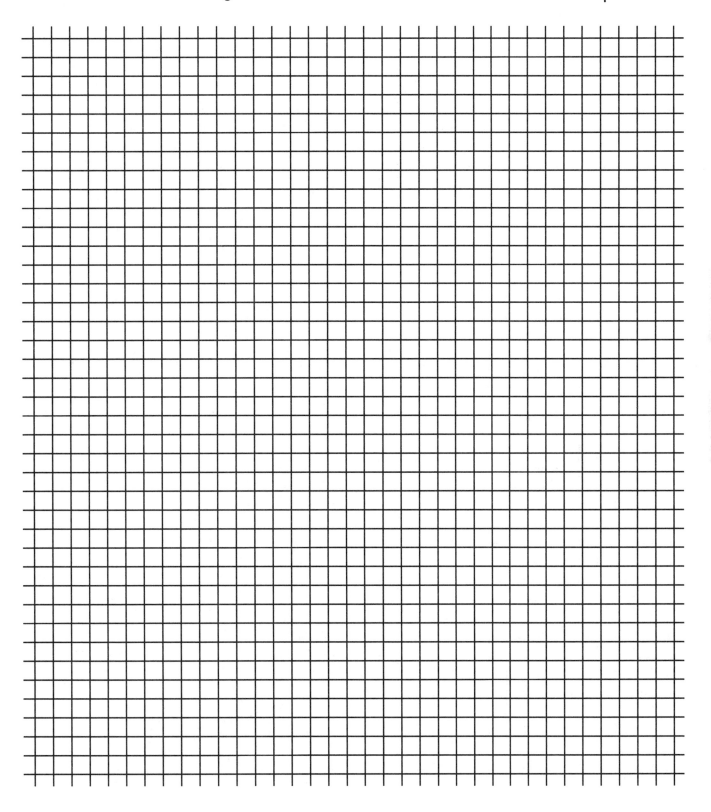

Multiplication Stretch

Topic
Multiplication and place value

Key Question
How can our base ten number system help simplify the conventional multiplication problem?

Learning Goals
Students will:
1. learn to multiply using the display multiplication method, and
2. become increasingly aware of the "ten-ness" of our numeration system and the role it plays in multiplication.

Guiding Documents
Project 2061 Benchmark
- *Multiply whole numbers mentally and on paper.*

NCTM Standards 2000*
- *Understand the place-value structure of the base-ten numeration system and be able to represent and compare whole numbers and decimals*
- *Understand the effects of multiplying and dividing numbers*
- *Develop and analyze algorithms for computing with fractions, decimals, and integers and develop fluency in their use*
- *Use the associative and commutative properties of addition and multiplication and the distributive property of multiplication over addition to simplify computations with integers, fractions, and decimals*
- *Select, apply, and translate among mathematical representations to solve problems*

Materials
Student sheets

Background Information
In display multiplication each partial product is computed and recorded separately as shown in the example on the first student sheet. Students are also asked to write an expression indicating which digits were used in finding the product.

Notice the absence of splitting the digits in the product, "carrying" and "indenting." It is the straightforward way in which this process of multiplication is carried out that appeals to many students.

Often students ask, "Can we do it this way all the time?" The answer is yes, particularly if it is more meaningful. The question itself is a commentary on what we do too often in teaching mathematics: prescribe too narrowly how students can do things.

In *Multiplication Stretch* the emphasis is on the concept of place value (ten-ness of our numeration system) and the concept of multiplication. Students must constantly keep in mind the place value associated with each of the digits being multiplied.

Management
1. This activity has students multiply two- and three-digit numbers in a way that they have probably never seen before. You may want to go over a few examples as a class so that all students are clear on the procedure before they begin.
2. Students can work on this activity independently or in small groups. There are advantages to both methods. You will have to decide which is better for your students.

Procedure
1. Hand out the student sheets and go over the instructions. Use the display method of multiplication to find the product of each of the problems given. You may want to do a few examples together as a class.
2. Have students work either alone or in small groups to complete the problems.
3. When all students have finished, close with a time of class discussion and sharing.

Discussion
1. How is this method different from regular multiplication? [It shows each of the partial products.]
2. Did you find the stretch method easier or more difficult than standard multiplication? Why?
3. What did you learn about multiplication from this activity?
4. What did you learn about place value from this activity?

Extension
Have students try multiplying four- and five-digit numbers with the stretch method.

* Reprinted with permission from *Principles and Standards for School Mathematics,* 2000 by the National Council of Teacher of Mathematics. all rights reserved.

Multiplication STRETCH

The example expands the by-products created through multiplication. Please use this method in finding the products in the following problems.

```
      43
  x   27
      21   = ( 7  x   3)
     280   = ( 7  x  40)
      60   = (20  x   3)
  +  800   = (20  x  40)
    1161
```

1.   35
   x 13
            = (x   )
            = (x   )
            = (x   )
   +   = (x   )

2.   72
   x 24
            = (x   )
            = (x   )
            = (x   )
   +   = (x   )

3.   59
   x 36
            = (x   )
            = (x   )
            = (x   )
   +   = (x   )

4.   48
   x 27
            = (x   )
            = (x   )
            = (x   )
   +   = (x   )

5.   87
   x 39
            = (x   )
            = (x   )
            = (x   )
   +   = (x   )

6.   63
   x 78
            = (x   )
            = (x   )
            = (x   )
   +   = (x   )

7.   92
   x 19
            = (x   )
            = (x   )
            = (x   )
   +   = (x   )

8.   394
   x 76
            = (x   )
            = (x   )
            = (x   )
            = (x   )
            = (x   )
   +   = (x   )

9.   487
   x 65
            = (x   )
            = (x   )
            = (x   )
            = (x   )
            = (x   )
   +   = (x   )

Square² Rules

Square Rules
1. Square the first number.
2. Double the cross-product of the first and second number.
3. Square the second number.

Topic
Square numbers

Learning Goals
Students will:
- practice squaring two-digit numbers to gain fluency with multiplication facts,
- extend thinking about meaning in conventional algorithms, and
- appreciate historical contributions to strategies for mental calculations.

Guiding Document
NCTM Standards 2000*
- *Identify and describe relationships between two quantities*
- *Identify, verify and express generalizations*
- *Use computational tools and strategies fluently*

Materials
Paper and pencil

Background Information
The story of Jakow Trachtenberg (adapted from *The Trachtenberg Speed System of Basic Mathematics* translated by Ann Cutler and Rudolph McShane, 1960) is of particular interest for at least three reasons. First of all, it is compelling to hear of the circumstance of being a prisoner in a Nazi concentration camp while he designed this speed system of mathematics. Secondly, during this time in history the necessity of speed in calculating numbers was real since calculators as we know them were not yet invented. And lastly, teachers as well as Professor Trachtenberg share a common goal of wanting to help students learn to calculate solutions to a variety of number problems rapidly "in their heads."

Professor Jakow Trachtenberg is the originator of a speedy system of calculation that appeals to students because it looks like magic but is based on arithmetic logic. This is his remarkable story of how this system evolved.

Trachtenberg was born in Russia near the end of the 19th century (1888) and grew up studying to become a mining engineer. Because of his dedication to pacifism, he became involved in organizing the Society of Good Samaritans and helped in caring for wounded soldiers in World War I. When Communism over took Russia in 1918, Trachtenberg strongly criticized the violence of killing and destruction. When he learned that he was slated to be murdered because of his influence and position, he fled to Germany.

There he met and married a beautiful and wealthy woman. He became quite well known for his knowledge and expertise in Russian industry, Russian history, and foreign languages. He also remained active in pushing Germany towards a future of peace. This position caused Trachtenberg to be seized by Hitler and placed in a concentration camp where he and many other Jews endured outrageous forms of punishment. To keep his sanity, he let his mind work with numbers arranging and rearranging them. It became a recreational outlet as he visualized gigantic numbers—to be combined or separated—devising short cuts for everything from addition and multiplication to algebra. He worked hard on simplifying this system by scribbling his ideas on scraps of paper and old envelopes.

When he learned he was to be executed after Easter in 1944, he entrusted his work to a fellow prisoner. Just before his scheduled execution, his wife bribed the prison guards and sold her jewels to arrange for her husband to be transferred to another prison camp. Nearly a year later in 1945, Trachtenberg again, aided by his wife, climbed through barbed wire fences as guards in watchtowers shot at him. Together they made it across the border to Switzerland where Trachtenberg recuperated in a Swiss camp for refugees. There he perfected his speed system of mathematics and in 1950 he founded The Mathematical Institute in Zurich—a school for children who were doing poorly in their schoolwork.

Square Rules is an adaptation of one of Trachtenberg's rules for fast mental calculations.

Procedure

1. Share the story of Trachtenberg. Explain that one of the "rules" that Trachtenberg thought would be helpful to students would be the ability to find the square of any two-digit number. Squaring a two-digit number equal to or less than 31 will produce a three digit answer; squaring a two-digit number greater than 31 will produce a four-digit answer.

2. Display the *Square Rules* on the overhead for all students to see.

Square Rules
1. Square the first number.
2. Double the cross-product of the first and second number.
3. Square the second number.

Explain that when Trachtenberg speaks of the first number, second number, and so forth, he treats each number in order from right to left. Thus, the first number is the digit furthest to the right (in the one's place) and the second number is its neighbor to the left (the number in the ten's place). Furthermore, you record only the answer. All other work is done in your head.

3. Work through the *Square Rules* with an example. For instance, 34.

---

1. Square the first number. (4 x 4 = 16) Record the 6 in the one's place in your answer and carry the 1.
2. Multiply the two digits in the problem and double the cross product. 2(3 x 4) = 24. Add the carried number (1). 24 + 1 = 25. Record the 5 in the ten's place in your answer and carry the 2.
3. Square the last number. (3 x 3 = 9) Add the carried number (2). 9 + 2 = 11. Record 11 in the hundred's and thousand's place.
4. The four-digit answer is 1156.

---

4. Ask students to work in pairs or small groups of three or four to try some problems of their own and to record them on large chart paper. Remind them that they are working only with two-digit numbers.

5. As a group, share some of the problems and their solutions. Suggest that students look for patterns or discoveries that are useful. Record questions and "ahas" along the way.

6. When students are feeling somewhat confident, invite them to discover Trachtenberg's special shortcut for two-digit numbers that end in 5. See *Five is a Special Case*.

Discussion

1. In your trial number squaring, what did you discover about the number of digits possible in an answer? [Squaring numbers less than 32 results in a three-digit answer. Numbers greater than 31 squared resulted in a four-digit answer.]

2. What patterns or insights did you find as you squared different numbers? [Answers will vary. Some may point out that with some numbers, if there is nothing carried in the one's place, a three-digit answer results. This is not always true. It also depends on the number in the ten's place.]

3. Explain how these special rules relate to the conventional way of multiplying a number by itself. Find the squares of the ones, the tens, and also double the cross-product in the partial products of the problem.

4. For *Five is a Special Case*, describe two ways to square a two-digit number that ends in a 5. [Most students will have no trouble describing that the digit in the one's place simply gets squared and recorded. Every two-digit number that ends in five, when squared results in a number that ends in 25. In generalizing what happens in the ten's place, students may describe that they multiplied the digit by itself and then added itself again ($n^2 + n$) or they may explain that they multiplied the digit in the ten's place by one more n ($n+1$).]

5. What other patterns or discoveries did you find? (Answers will vary.)

* Reprinted with permission from *Principles and Standards for School Mathematics*, 2000 by the National Council of Teachers of Mathematics. All rights reserved.

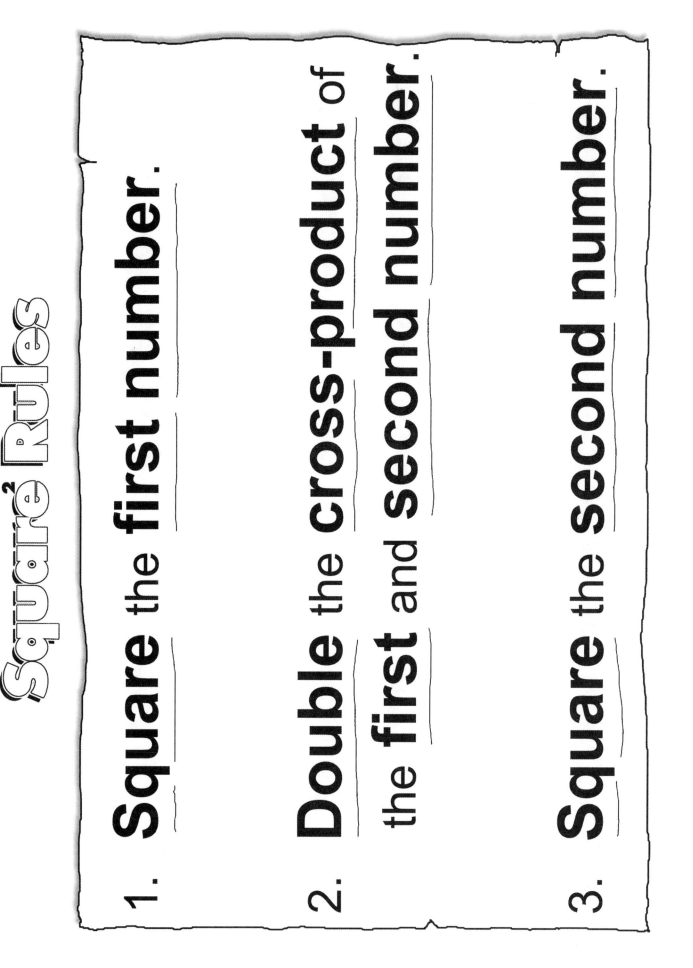

Square² Rules

1. **Square** the **first number**.

2. **Double** the **cross-product** of the **first** and **second number**.

3. **Square** the **second number**.

Square² Rules
Five is a Special Case

Challenge: Can you discover Trachtenberg's special case rule for squaring a two-digit number that ends in a five?

Try these problems and look for a pattern that you can generalize!

15	25	35	45	55
x 15	x 25	x 35	x 45	x 55

	65	75	85	95
	x 65	x 75	x 85	x 95

Explain and describe short cuts, patterns, and discoveries you found.

Just the Answers!

Topic
Multiplication

Learning Goals
Students will:
- experience alternate methods of calculation, and
- examine connections to conventional algorithms for multiplication.

Guiding Document
NCTM Standards 2000*
- *Understand numbers, ways of representing numbers, relationships among numbers, and number systems*
- *Use computational tools and strategies fluently and estimate appropriately*

Materials
Just the Answer recording sheet

Background Information
The biographical story of Jakow Trachtenberg is included with the lesson *Square Rules*. Speedy ways of multiplying any number by 11 or 12 is another contribution from this gifted man.

Procedure
1. Read the story about Jakow Trachtenberg and describe his focus on making calculations fast and easy for school students. Tell students that they are going to look at short cuts for multiplying any number by 11 or 12 and be able to write down just the answer. All work is done in your head.
2. Describe the rules for multiplying by 11, making sure that students understand that only the answer is written and that work begins in the units place and moves to the left.

Multiplying by Eleven

Rules with example: 243 x 11 = 2673

1. Bring down the first number into your answer. 3
2. Add the next number to its neighbor on the right. Record. 4 + 3 = 7
3. Continue adding each number to its right hand neighbor and record as part of the answer. 2 + 4 = 6
4. Bring down the last number into your answer. 2
5. In cases where the two numbers added result in a two-digit number, write down the units and carry the tens to the next number.

---

3. Turn to your neighbor and explain the rules as you understand them. Compare your answers to this problem: 1326 x 11 = 14,586
4. Distribute *Just the Answer!* Complete the problems writing only the answers.

136 x 11 = 1496	738 x 11 = 8118
178 x 11 = 1958	952 x 11 = 10,472
224 x 11 = 2464	2735 x 11 = 30,085
532 x 11 = 5852	43523 x 11 = 478,753
596 x 11 = 6556	3821547 x 11 = 42,037,017

5. After students have completed the problems, suggest that they compare their answers with a neighbor. Work out places where there are differences.
6. Tell the students to be prepared to share their work and their thinking about how and why the Trachtenberg rules work.

7. Explain that there is a similar rule for multiplying any number by 12. Explain the rules.

Multiplying by Twelve

Rules with example: 234 x 12 = 2808

1. Double the first number and bring it down into your answer. (4 x 2 = 8)
2. Double each number in turn and add to its neighbor on the right. Record. (3 x 2) + 4 = 10. Bring down the 0 and carry the 1.
3. Continue doubling each number and adding its neighbor on the right. (2 x 2) + 3 + 1= 8
4. Bring down the last number. 2

---

8. Invite students to try *Double the Fun!* (Answers are included here)

$$123 \times 12 = 1476$$
$$254 \times 12 = 3048$$
$$829 \times 12 = 9948$$
$$4136 \times 12 = 49,632$$
$$35461 \times 12 = 425,532$$

Discussion
1. Describe how the are "rules" for 11 and 12 the same or different.
2. How does this method compare to using the conventional algorithm?
3. What are the advantages and disadvantages of using the Trachtenberg rules?
4. How do these rules incorporate ideas about place value? [because you are really multiplying by one and then by ten and adding the partial products]
5. Since addition is commutative, could you add to the neighbor on the left instead of the one on the right? Explain. [With 11s, it would make no difference. With 12s, since doubling each digit is important, you would have to add the digit on the right to double the digit on the left. Therefore it becomes more efficient to double and then add to the right hand neighbor.]
6. What is your thinking about a rule for multiplying a number by 13?

* Reprinted with permission from *Principles and Standards for School Mathematics,* 2000 by the National Council of Teachers of Mathematics. All rights reserved.

Just the Answers!

Use the speed system of basic multiplication to multiply these numbers by 11.

Rules with example: 243 x 11 = 2 6 7 3

243

2673

1. Bring down the first number into your answer. 3
2. Add the next number to its neighbor on the right. Record. 4 + 3 = 7
3. Continue adding each number to its right hand neighbor and record as part of the answer. 2 + 4 = 6
4. Bring down the last number into your answer. 2
5. In cases where the two numbers added result in a two-digit number, write down the units and carry the tens to the next number.

Problem	Answer	Problem	Answer
136 x 11 =		738 x 11 =	
178 x 11 =		952 x 11 =	
224 x 11 =		735 x 11=	
532 x 11 =		43523 x 11 =	
596 x 11 =		3821547 x 11=	

Think like Trachtenberg. How do you think he discovered this speed system?
Share your ideas, thinking, and planning on the back side of this paper.

Use the speed system of basic multiplication to multiply these numbers by 12.

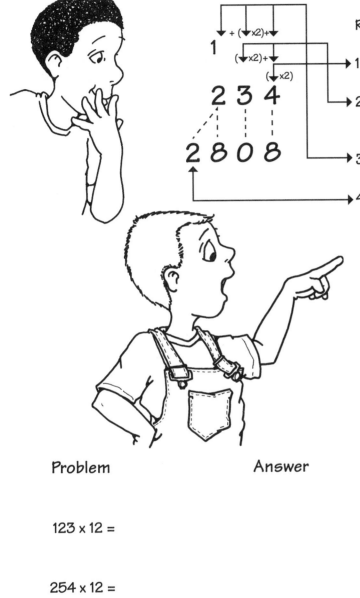

Rules with example: 234 x 12 = 2808

1. Double the first number and bring it down into your answer. (4 x 2 = 8)

2 Double each number in turn and add to its neighbor on the right. Record. (3 x 2) + 4 = 10. Bring down the 0 and carry the 1.

3. Continue doubling each number and adding its neighbor on the right. (2 x 2) + 3 + 1= 8

4. Bring down the last number. 2

Problem   Answer

123 x 12 =

254 x 12 =

829 x 12 =

4136 x 12 =

35461 x 12 =

Quick Sticks and Lattice Multiplication

Topic
Multiplication

Learning Goal
Students will analyze and evaluate the mathematical thinking of John Napier, a Scottish mathematician, and his strategies for rapid calculation of multiplication problems using a set of wooden rods.

Guiding Document
NCTM Standards 2000*
- *Select appropriate methods and tools for computing with whole numbers from among mental computation, estimation, calculators and paper and pencil according to the context and nature of the computation and use the selected method or tool*
- *Analyze and evaluate the mathematical thinking and strategies of others*

Materials
For each student:
10 flat wooden sticks such as craft sticks
highlighter marker
pattern strips for Napier's Rods (Quick Sticks)
white glue

Background Information
John Napier, a 16th century Scottish mathematician, is credited with the design of a set of wooden rods marked with numbers to assist with rapid multiplication of numbers.

In this lesson students construct a facsimile of these rods and use them to calculate answers to multiplication problems. For some students, the process is simplified because the cross products are completed first and then the answer is gathered by adding along the diagonals. This process connects well to the conventional algorithm and to the lattice method of multiplication.

Using the quick sticks: Select sticks to match the problem. For 76 x 4, select the 7 stick and the 6 stick. Align sticks 7 and 6 side by side. Use Index stick to find the 4th row. Add along diagonals moving right to left.

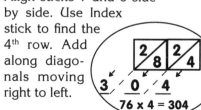 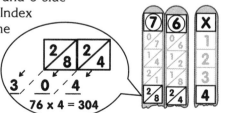

76 x 4 = 304

Management
1. Using glue is messy. Advise students that "less is better."

Procedure
1. Distribute 11 flat sticks (an index rod (x) and 10 number rods for 0–9) and one copy of *Quick Sticks* to each student.
2. Cut and glue each strip to a wooden stick and set aside to dry.
3. Share the historical contributions of John Napier and suggest that his thinking and strategies could be helpful to us today. (See AIMS publication *Historical Connections, Vol. I.*)
4. Introduce Lattice Multiplication as a method of multiplication used long ago by the Hindus and could possibly have been the source of the idea for Napier's rods.
5. Use the sample problems and explain that the problems are set up "outside the box" on the upper and right sides. For example, 46 x 3 looks like this.

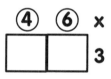

Two columns of unit boxes are needed for 46 and one row of unit boxes for 3. For larger numbers, such as 142 x 56, three columns of unit boxes are needed and two rows of unit boxes.

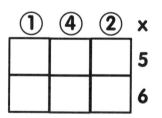

In general, you need a column or a row of unit boxes for every number to be multiplied. The number of unit boxes needed is equal to the number of digits in one number times the number of digits in the other number.

6. Prepare each unit box formed by drawing a diagonal in each box to separate the place values of the cross products.

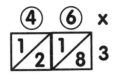

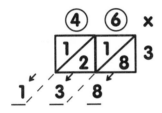

Record the product of each pair of factors in the respective unit boxes. Each box is the product of two factors with only a two-digit response.

The final step is the result of adding along the diagonals beginning in the lower right corner and moving to the left.

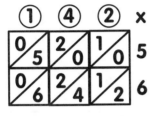

Example A

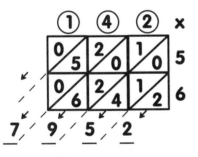

Example B

Encourage students to try problems of their own. Suggest that they compare the results and the process of this method to the traditional algorithm.

7. When the *Quick Sticks* are dry, try doing the same problems using the "sticks" by lining them up and adding along the diagonals.

8. Distribute *Quick Sticks and Lattice Multiplication.* Use the sticks to solve the problems.

Discussion
1. How is Lattice Multiplication similar to using Napier's rods or Quick Sticks? [Both show partial products in boxes marked by a diagonal.]
2. How would you compare Quick Sticks, Lattice Multiplication, and the conventional method of multiplying? How are they alike? How are they different?
3. Which method is faster (for you)?
4. How essential is knowing the multiplication facts for each method?

* Reprinted with permission from *Principles and Standards for School Mathematics,* 2000 by the National Council of Teachers of Mathematics. All rights reserved.

Quick Sticks and Lattice Multiplication

Index

1. Highlight in yellow or a bright color the number heading each column and the numbers on the Index rod (x).
2. Cut out each strip and glue to a wooden craft stick.

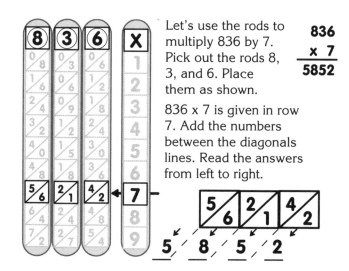

Let's use the rods to multiply 836 by 7. Pick out the rods 8, 3, and 6. Place them as shown.

$$\begin{array}{r} 836 \\ \times\ 7 \\ \hline 5852 \end{array}$$

836 x 7 is given in row 7. Add the numbers between the diagonals lines. Read the answers from left to right.

5 8 5 2

Now let's multiply a three digit number by a two digit number. Pick out the rods 6, 7, and 4. Place them as shown.

674 x 48

$$\begin{array}{r} 674 \\ \times\ 48 \end{array}$$

Read from row 8   5392
Read from row 4   2696
$$\overline{32{,}352}$$

A method of multiplication used by the early Hindus, is called lattice multiplication. A lattice diagram is drawn and additions are performed diagonally.

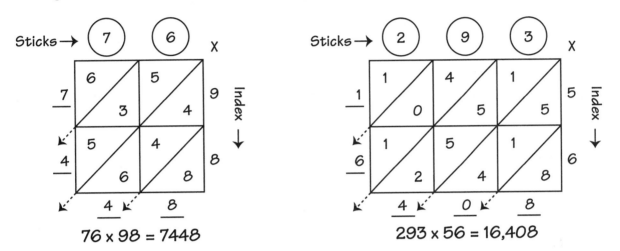

$$76 \times 98 = 7448$$

$$293 \times 56 = 16,408$$

Use lattice multiplication and Quick Sticks to solve these problems.

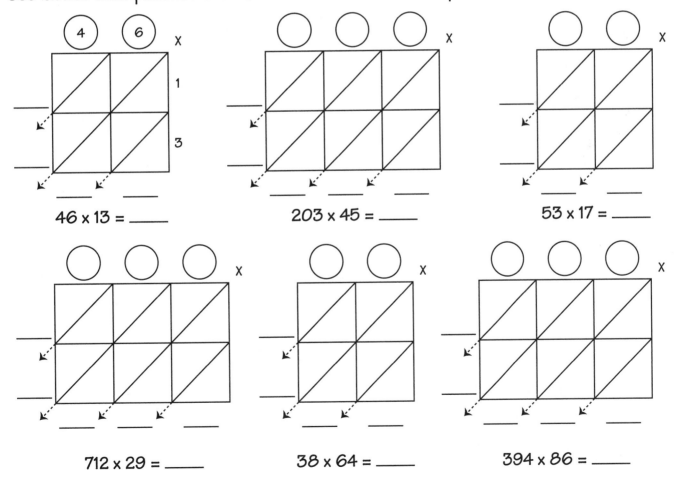

$$46 \times 13 = \underline{\qquad}$$

$$203 \times 45 = \underline{\qquad}$$

$$53 \times 17 = \underline{\qquad}$$

$$712 \times 29 = \underline{\qquad}$$

$$38 \times 64 = \underline{\qquad}$$

$$394 \times 86 = \underline{\qquad}$$

Learning Goal
Students will explore the patterns in the multiplication table.

Materials
For each group of four students:
    72 index cards, 3" x 5"
    4 colored markers
    scissors
    copy of the multiplication table (included)

Management
1. Colored markers for each group should match.
2. Students may need to draw a line underneath the numbers nine and six to distinguish them from each other.

Procedure
Part One—Making the Cards
1. Tell the students to cut the multiplication table provided into four equal sections marked by bold lines and distribute one section to each student in the group.
2. Direct students to fold each index card in half and cut it along the fold to produce a total of 144 squares.
3. Have each student copy the numbers from their section of the multiplication table onto 36 cards.

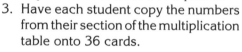

Part Two—Playing the Game
1. Explain that the object of the game is to build the multiplication table with their cards.
2. Direct one student in each group to shuffle the set of 144 cards and deal 36 cards to each person in the group.
3. Explain that play begins with the student to the left of the dealer who places one of his/her cards face up on the table.
4. Play continues in a clockwise fashion as each student plays a card that touches a card already played on the table. The card may share an edge or a diagonal corner. If a player cannot play a card, move on to the next player.

DEALER ➡

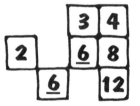

5. Each player must decide on his/her own which card to play and must explain this choice of card and where it fits best. Teammates may encourage but not choose for the player.
6. After the table is completed, encourage team discussions of strategies and patterns that are helpful.

Winning Patterns

1	2	3	4	5	6	7	8	9	10	11	12
2	4	6	8	10	12	14	16	18	20	22	24
3	6	9	12	15	18	21	24	27	30	33	36
4	8	12	16	20	24	28	32	36	40	44	48
5	10	15	20	25	30	35	40	45	50	55	60
6	12	18	24	30	36	42	48	54	60	66	72
7	14	21	28	35	42	49	56	63	70	77	84
8	16	24	32	40	48	56	64	72	80	88	96
9	18	27	36	45	54	63	72	81	90	99	108
10	20	30	40	50	60	70	80	90	100	110	120
11	22	33	44	55	66	77	88	99	110	121	132
12	24	36	48	60	72	84	96	108	120	132	144

Russian Peasant Method

Topic
Multiplication

Learning Goals
Students will:
- experience alternate methods of calculation by multiplication, and
- appreciate a historical context for alternate processes of calculation.

Guiding Document
*NCTM Standards 2000**
- *Use computational tools and strategies fluently and estimate appropriately*
- *Understand various meanings of multiplication and division*

Background Information
This experience is intended to provide an interesting look at an alternative method of multiplication. It will appear to be a trick to most elementary students and teachers. The explanation regarding why it works is not the focus of the experience but rather that there are many ways to do multiplication and some ways are more efficient than others. For those students who might appreciate further explanation regarding why it works, the following description may be helpful.

The logic of the method is based on the translation of one of the numbers in the problem into powers of two and using the other number as the multiplier. Let's use the sample problem 18 x 25. This could be thought of as 18 sets of 25. If we express 18 as the sum of powers of two, we get 2 + 16. When we add the products of 2 x 25 and 16 x 25, we get the answer to the problem, 450. Another way to visualize this is to create a third column of the powers of 2 and align them with the numbers generated in the solution. The rows that are **not** crossed out are the second row and the last row that align themselves with 2 and 16. Each is multiplied by 25, and the resulting sum is 450.

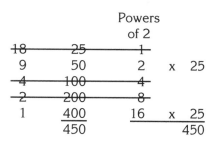

		Powers of 2		
~~18~~	~~25~~	~~1~~		
9	50	2	x	25
~~4~~	~~100~~	~~4~~		
~~2~~	~~200~~	~~8~~		
1	400	16	x	25
	450			450

In the 1800s peasants in a remote area of Russia were observed using this remarkable and unusual method of multiplication.

Procedure
1. Try multiplying 18 x 25 the Russian Peasant way.
2. Write the problem so that one number heads one column and the other heads a second column.

$$18 \quad x \quad 25$$

3. Halve each number in the left column disregarding remainders, and double the numbers in the right column.

Halve this column (disregard remainders)		Double this column
18	x	25
9		50
4		100
2		200
1		400

4. Cross out all the rows of numbers that begin with an even number on the left; then add up the remaining numbers on the right.

~~18~~	~~25~~
9	50
~~4~~	~~100~~
~~2~~	~~200~~
1	400
	450

* Reprinted with permission from *Principles and Standards for School Mathematics,* 2000 by the National Council of Teachers of Mathematics. All rights reserved.

Russian Peasant Method

Try these problems on your own.

12 x 25 =			20 x 34 =		
Left Column (halve)		**Right Column** (double)	**Left Column** (halve)		**Right Column** (double)
12	x	25	20	x	34

Try a two-digit multiplication problem of your own design.

Check your work against the traditional method. What similarities can you find? Which method do you prefer? Why?

What happens if the number in the left-hand column is an odd number?

Try this problem 17 x 10.

Building Conceptual Understanding
Division

Concrete/Manipulative Level

At this level students partition or separate a set of objects into two or more equal parts to experience the basic operation in a concrete way.

Using countable objects

Using countable objects, such as counting bears or seashells, students begin with a set of objects and fair share them into *equal* sets. By counting the number of objects in *one* fair share, students are able to determine the quotient.

Using measurable objects

Using measurable objects such small ceramic tiles or paperclips, students construct a train or a rectangular array and then break into two or more equal parts or fair shares.

Counted

Measured

Representational/Pictorial Level

At the representational level, sometimes called the connecting stage, objects and actions are represented or depicted by pictures or diagrams. For division, a set of objects is pictured and arrows indicate that the set has been fair shared into two or more equal sets.

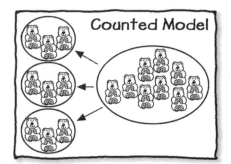

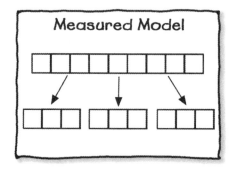

Abstract/Symbolic Level

At this level numerals, symbols, and relationship signs are used to represent the objects and actions of the concrete and representative levels. At the symbolic level, the number of objects to be fair shared represents the dividend, the number of fair shares is the divisor, and the number of objects in one fair share is the quotient. The ÷ or ⌐ signs represents the process of dividing and the = sign shows that no objects have been gained or lost in the transaction.

Fifteen divided by three is five

$$15 \div 3 = 5 \qquad \text{or} \quad 3\overline{)15}^{\,5}$$

One fair share has five objects if three fair shares have 15 objects.

Topic
Division by repeated subtraction or "clustered" subtraction

Process
In repeated subtraction, the divisor is repeatedly separated from the dividend until it can no longer be subtracted. The number of separations is totaled and the remainder is added as appropriate. In clustered subtraction, multiples of divisors are separated in easy, large pieces. For example, 10 times the divisor is an easy part to remove, thus reducing the dividend for easier factoring. Then the partial quotients are added.

Repeated Subtraction

$$13\overline{)246} = 18 \text{ R } 12$$

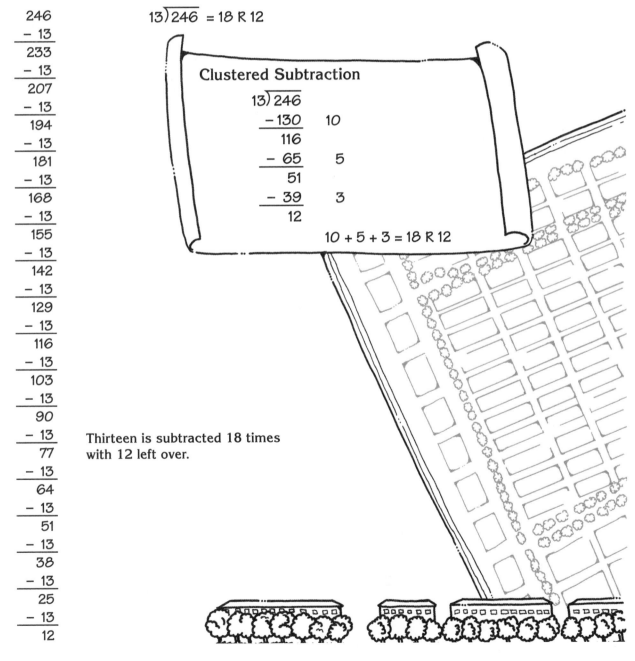

```
  246
 - 13
  233
 - 13
  207
 - 13
  194
 - 13
  181
 - 13
  168
 - 13
  155
 - 13
  142
 - 13
  129
 - 13
  116
 - 13
  103
 - 13
   90
 - 13
   77
 - 13
   64
 - 13
   51
 - 13
   38
 - 13
   25
 - 13
   12
```

Clustered Subtraction

$$13\overline{)246}$$

```
13)246
 -130    10
  116
 - 65     5
   51
 - 39     3
   12
```

10 + 5 + 3 = 18 R 12

Thirteen is subtracted 18 times with 12 left over.

Topic
Division as the process of fair shares

Learning Goals
Students will:
- use a small set of objects such as plastic counting bears to create fair shares as a way of modeling the process of division, and
- experience multiplication and division as inverse operations.

Guiding Document
NCTM Standard 2000*
- *Understand meanings of operations and how they relate to one another*

Materials
For each pair of students:
> 20 Teddy Bear counters or other small countable objects
> one paper raft
> six paper boats
> one set of *Summer Camp Cards*

Management
1. Duplicate the raft and boat pages. Cut out six boats and one raft for each team of two students.
2. Duplicate a set of *Summer Camp Cards* for each pair of students.

Background
This activity is intended to introduce the idea of division as the creation of fair shares. Only after students use the manipulatives and show some sense of understanding should a paper/pencil record be introduced.

Each action at the concrete level can be traced through a corresponding written record. If the problem is to divide 15 bears into 3 boats, the first step is to place 15 bears on the raft represented by an outline of a box. The box gradually becomes the division sign.

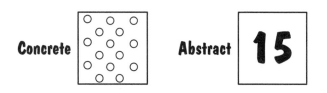

Step One: 15 bears on the raft

The second step is to determine how many fair shares are to be formed. In this case, three to represent the three boats. The three is recorded to the left of the box.

Step Two: 3 boats pull up next to the raft

The third step is to make three fair shares. The five is written above the box showing that there are 5 bears in each fair share or each boat.

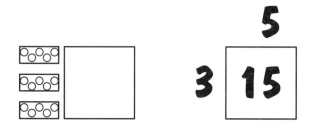

Step Three: Make fair shares.

Procedure
1. Distribute rafts, boats, and counting bears to each pair of students.
2. Describe the context of the story and propose the first problem to be solved.

Betty Bears and Teddy Bears are having a great time at summer camp in the mountains. Boats are used to travel back and forth from a raft anchored in the middle of the lake to the shore. Every time the boats are used to carry bears, each boat must carry the same number of bears for that trip. The places where the boats arrive at the raft are marked with buoys. Use the raft, some boats and bears to solve some problems.

3. Pose a sample problem to students. Six (6) bears are on the raft and three (3) boats come to take them ashore. How many bears will be in each boat?

4. Ask students to model the problem as it is read aloud. Guide their work by telling them to place 6 bears on the raft and to move 3 boats into the lanes marked by buoys.

5. Ask them to fair share the bears one at a time into each boat so that no boat has more bears than any other boat.

6. Ask them to explain how many bears are in each boat.

7. Distribute one page of six *Summer Camp Cards* to each pair of students.

8. Tell students to cut apart the cards and place them face down in a stack in front of them.

9. Explain that they will take turns reading the problem on a card to their partner and the partner models the problem with the bears, boats, and raft.

10. After pairs of students have completed experiencing the process of division as fair shares, connect the process to the symbolic record of division by explaining the three parts of the conventional division algorithm. The bears on the raft represent the dividend, the boats represent the divisor, and the bears in one boat represent the quotient.

11. Distribute *Fair Shares: On My Own* and ask students to show what they know by using the bears, boats, and rafts to complete the four problems described and to show their work and their thinking.

Discussion

1. What does it mean to make a fair share with bears, boats, and a raft? [All boats have the same number of bears.]

2. How do the bears on the raft connect to doing division? [The bears on the raft represent the number under the division sign.]

3. What part of the division problem represents the boats? [the divisor]

4. The number of bears in one boat represents the quotient. What is another name for quotient? [the answer to a division problem]

5. What could we do if there were 16 bears and three boats? (Discussion should include what to do with remainders.)

* Reprinted with permission from *Principles and Standards for School Mathematics,* 2000 by the National Council of Teachers of Mathematics. All rights reserved.

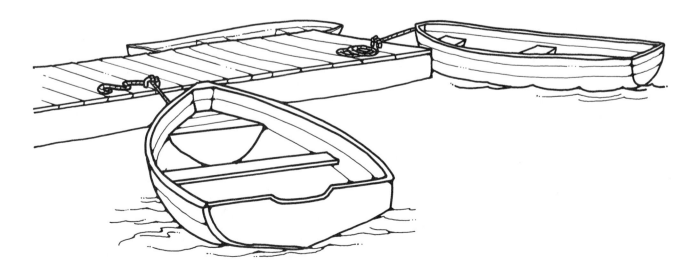

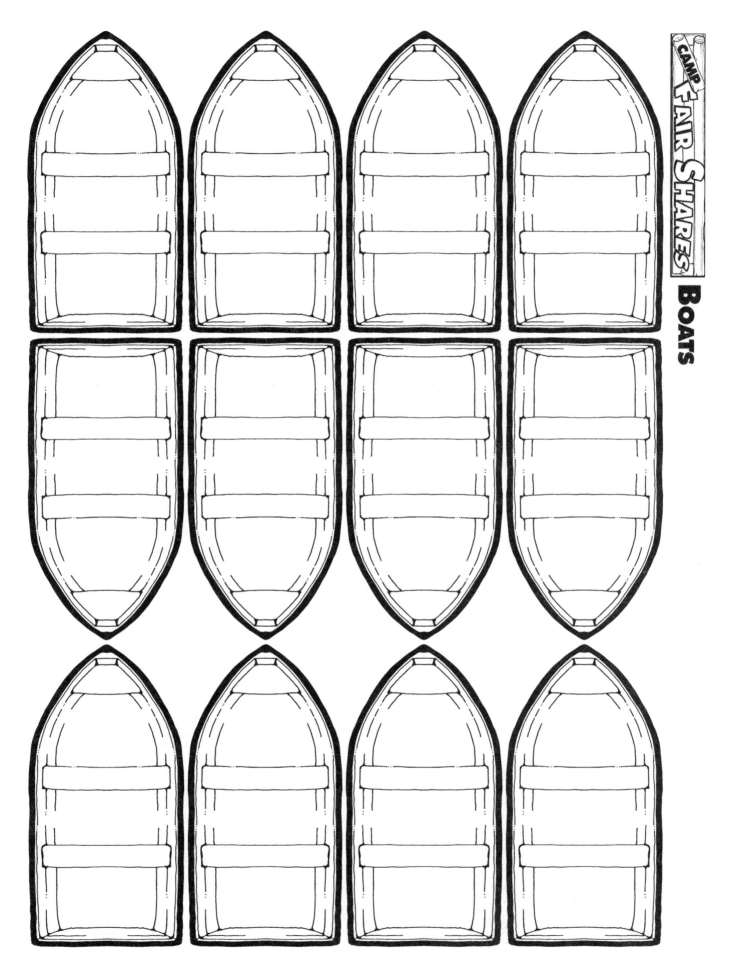

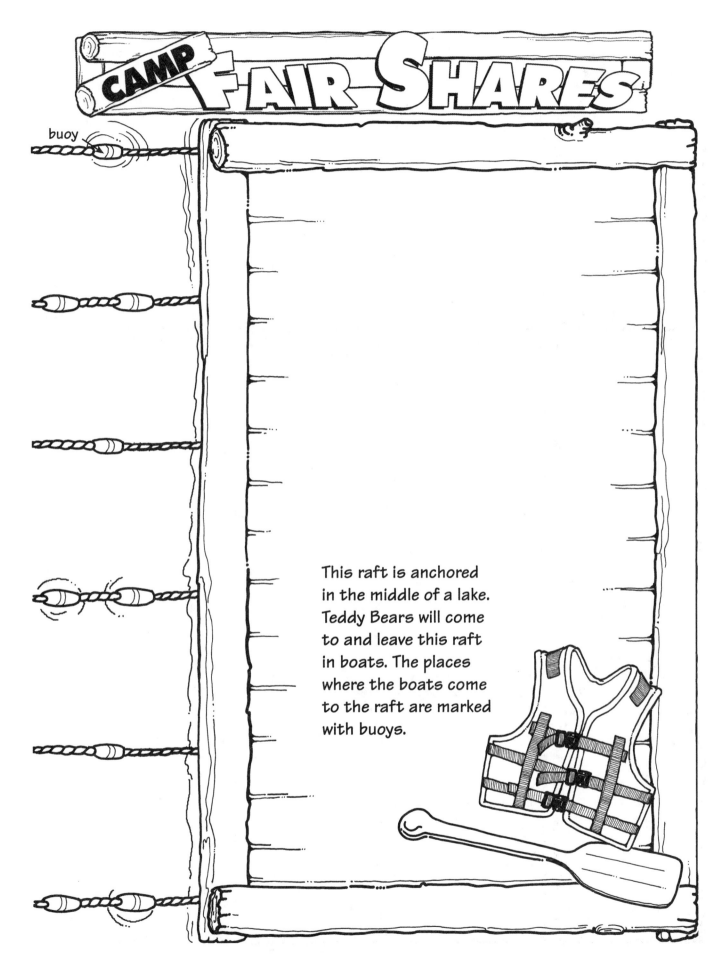

buoy

This raft is anchored in the middle of a lake. Teddy Bears will come to and leave this raft in boats. The places where the boats come to the raft are marked with buoys.

CAMP Fair Shares

6 Teddy Bears are on the raft. They use 3 boats to come ashore. How many will be in each boat?

15 Betty Bears want to use 3 boats to come ashore. How many must get into each boat?

3 boats are carrying 9 Betty Bears to shore. How many are in each boat?

4 boats are bringing 12 Teddy Bears to shore. How many are in each boat?

8 Teddy Bears are coming ashore on 4 boats. How many are in each boat?

12 Teddy Bears are waiting to come ashore on 2 boats. How many will be in each boat?

16 Betty Bears are getting into 4 boats. How many will be in each boat?

20 Betty Bears plan to come ashore in 4 boats. How many will get into each boat?

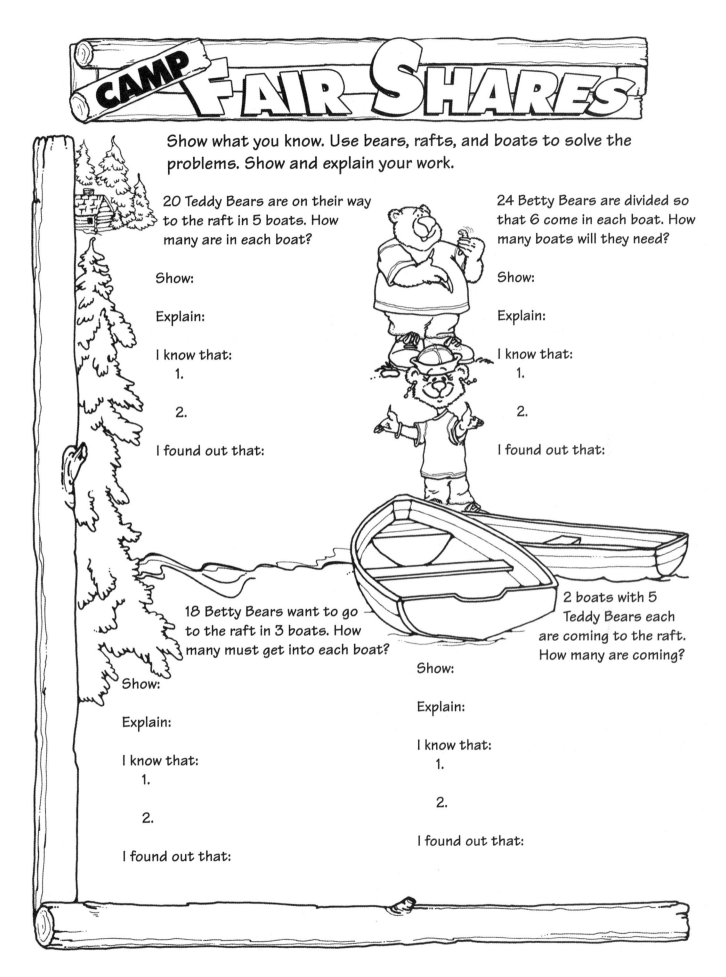

CAMP Fair Shares

Show what you know. Use bears, rafts, and boats to solve the problems. Show and explain your work.

20 Teddy Bears are on their way to the raft in 5 boats. How many are in each boat?

Show:

Explain:

I know that:
1.

2.

I found out that:

24 Betty Bears are divided so that 6 come in each boat. How many boats will they need?

Show:

Explain:

I know that:
1.

2.

I found out that:

18 Betty Bears want to go to the raft in 3 boats. How many must get into each boat?

Show:

Explain:

I know that:
1.

2.

I found out that:

2 boats with 5 Teddy Bears each are coming to the raft. How many are coming?

Show:

Explain:

I know that:
1.

2.

I found out that:

Topic
Whole number operations: Division

Learning Goals
Students will:
- understand the concept of division as a process of making fair shares,
- connect what it means to "do" division at the manipulative level with the "writing" about division at the abstract level, and
- practice fair sharing with a concrete model.

Guiding Document
NCTM Standards 2000*
- *Understand various meanings of multiplication and division*
- *Understand the effects of multiplying and dividing whole numbers*

Materials
8-10 cloth or paper bags (see *Management 1*)
Countable objects (see *Management 2* and *3*)
16 dice (cubicle random generators)
Boxing Rings mats
1 bell
Boxing Score sheets

Management
1. Prepare as many boxing bags as there are groups of students. Place small countable objects in each bag.
2. Examples of countable objects are buttons, washers, cubes, counting bears, plastic chips, dried beans, pencil erasers, tiny sea shells, and so forth.
3. The grade level at which the students are operating determines the number of objects in each bag. Perhaps 30-50 in some bags and more in others, up to 80, would be reasonable. Numbers significantly higher simply make the process tedious and lose sight of the focus of the experience—division is a process of partitioning into fair or equal shares.
4. There will be equally as many "boxing rounds" as there are bags of objects and groups of students. When the bell rings, boxing bags are exchanged with another group.

Procedure
1. Explain to students that they will be involved in several boxing rounds during which they will receive a boxing bag with objects to be shared equally or distributed into matching boxes so that the number in each box matches all the other boxes. Each round begins and ends at the sound of the bell.
2. Explain that the contents of the boxing bag will be spilled into the center of the *Boxing Ring*. The die is then rolled to determine how many boxes are placed next to the ring. If a 1 is rolled, it becomes a 7. All other numbers are read at their face value. Then all of the objects are equally shared into the matching boxes in one to one correspondence. Any remaining objects are left in the boxing ring.
3. Demonstrate to the class, the process of setting up the boxing ring, placing the objects in the ring, and fair sharing them into matching boxes next to the ring.
4. Make sure students understand that each group is responsible for recording their boxing scores on the scorecard. Each member of the team assumes the role of boxing referee for at least one round. It is the responsibility of the boxing referee to check the recording of the boxing round and sign off on it when he or she has approved it.
5. Demonstrate this process of scoring each round. Each round is numbered and has a boxing ring that represents the number of objects in the bag. Next to the box is a place to record the number

of matching boxes that were filled. The space above the boxing ring indicates the number of objects in each matching box and the number of leftovers, if appropriate.

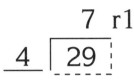

6. Distribute to each group one bag of objects, one die, one *Boxing Ring* mat, and a *Boxing Score* sheet.
7. Ring the bell and begin the rotations.

Discussion
1. If division is the process of separating a set of objects into two or more equal parts, how are the boxing rings and matching boxes like the process of division? [The boxing ring represents the set of objects to be separated into equal sets. The matching boxes outside the ring provide a place for equal sets to be partitioned.]
2. Three parts of the division problem are the dividend, the divisor, and the quotient. Explain each term as it relates to the boxing matches. [The number of objects in the boxing ring represents the dividend. The number of matching boxes outside the ring is the divisor. The number of objects in each equal box is the quotient.]

Evaluation
At the concrete level:
    Ask students to use objects to show what this problem means and how to do it.

$$4\overline{)29}$$

At the representational level:
    Ask students to draw a picture or diagram that shows what this problem means.

$$3\overline{)16}$$

At the abstract level:
    Ask students to write the problem that best matches this picture.

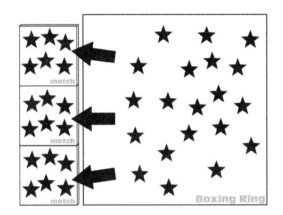

At an applied level:
    Ask students to write a short word problem that requires the division operation.

Evidence of Learning
1. Look for action at the concrete level that matches the problem. Students should begin with a large set of objects matching the dividend and show a move to create equal shares represented by the divisor.
2. Listen for appropriate use of mathematical language that is not just parroted. Listen for genuine use of words such as dividing by, divisor, equal sets or groups, fair shares, and so forth.
3. Look for pictures and symbols appropriate to the problem.
4. Listen for explanations and situations that connect to making equal sets from a larger set.

*   Reprinted with permission from *Principles and Standards for School Mathematics,* 2000 by the National Council of Teachers of Mathematics. All rights reserved.

BOXING BAGS & MATCHES

MATCHES MATCHES

Boxing Ring

Match Boxes

match

match

match

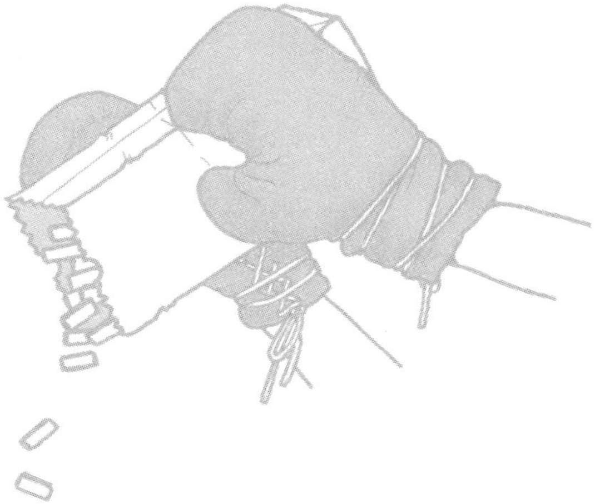

match

match

match

match

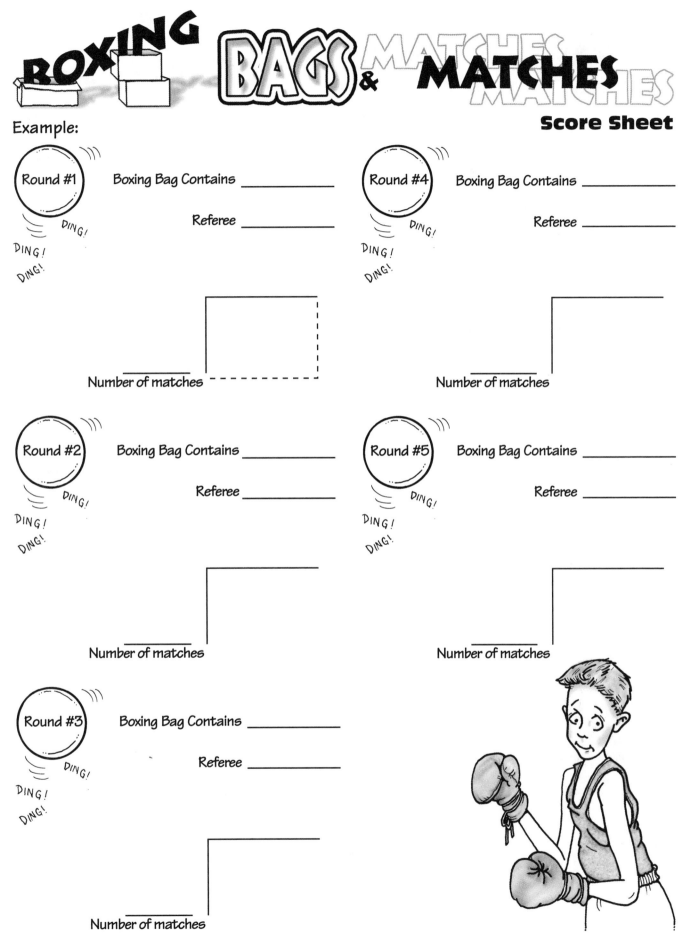

BOXING BAGS & MATCHES

Score Sheet

Example:

Round #1   DING! DING! DING!

Boxing Bag Contains _____

Referee _____

_____ Number of matches

Round #2   DING! DING! DING!

Boxing Bag Contains _____

Referee _____

_____ Number of matches

Round #3   DING! DING! DING!

Boxing Bag Contains _____

Referee _____

_____ Number of matches

Round #4   DING! DING! DING!

Boxing Bag Contains _____

Referee _____

_____ Number of matches

Round #5   DING! DING! DING!

Boxing Bag Contains _____

Referee _____

_____ Number of matches

BOXING BAGS & MATCHES MATCHES

MATCHES
MATCHES

Round #6

Boxing Bag Contains _____

Referee _____

DING!
DING!
DING!

_____ |_____

_____
Number of matches

Round #7

Boxing Bag Contains _____

Referee _____

DING!
DING!
DING!

_____ |_____

_____
Number of matches

Round #8

Boxing Bag Contains _____

Referee _____

DING!
DING!
DING!

_____ |_____

_____
Number of matches

PACK-10 Trading Centers

Topic
Division of whole numbers

Learning Goals
Students will:
- understand meaning of multi-digit division in the base ten number system,
- practice division at the concrete level with a manipulative, and
- experience the process of division in a playful and meaningful context.

Guiding Document
NCTM Standard 2000*
- *Understand the effects of multiply and dividing whole numbers*

Materials
For each small group or pair of students:
one set of AIMS Base Ten Manipulatives
one laminated loading dock
6-7 trucks
Bill of Lading forms

Management
1. The AIMS Base Ten Manipulatives are effective for modeling each of the basic operations. Each set consists of 70 cubes (ones), 40 longs (tens), and 10 flats (hundreds). They can be found in the AIMS catalog, Item #4008.
2. The teacher will need to prepare ahead of time a pre-printed mat that serves as a loading dock for each set of students. Laminating these mats increases their durability and length of use.

Background
The AIMS Base Ten Manipulatives provide the opportunity for students to do division at the concrete level and thus establish a basis for recording each action and connecting it to the abstract written form. The primary focus of this activity is to model the process of division and connect it to the written record of the conventional division algorithm.

Procedure
1. Distribute a set of base-ten manipulatives to each small working group.
2. As students explore the materials, confirm that they understand the "ten-for-one" idea. Ten cubes = one long; 10 longs = one flat.
3. Set the stage for *Pack-10 Trading Centers* by suggesting that the cubes, longs, and flats represent cartons, stacks, and pallets. The cartons are single units, while ten cartons make a stack (10 cubes) and 10 stacks fill a pallet (100 cubes). Explain to the students that the Bug-Lite Trucking Company wants them to use the loading dock to assemble the materials and then pack them equally into the allocated containers on the trucks that pull up to the loading dock. It is important that the loading be done with the most efficient number of packing materials.
4. Make sure students understand the "big idea" and then try a sample problem.
   Problem #1: "Place on the loading dock the equivalent of 47 cartons using the fewest number of packing materials possible." In this case, watch to see that they use 4 stacks and 7 cartons rather than 47 cartons. "Move 3 Bug-Lite Trucks up to the loading dock and load each truck with equal sets of materials."

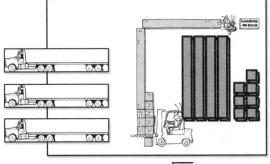

This matches the problem $3\overline{)47}$ where 47 is the dividend and 3 is the divisor. Discuss with students how many cartons are on each truck and how many are left on the loading dock. The number of cartons in each truck matches the quotient and the cartons left on the loading dock corresponds to a remainder.

5. Distribute *Bill of Lading* forms and have the students practice "doing" division, picturing it and writing it in symbolic form. Have students use their loading dock mats and trucks to solve the following problems.

   1. $47 \div 3$
   2. $52 \div 3$
   3. $71 \div 4$
   4. $84 \div 3$
   5. $127 \div 3$
   6. $212 \div 3$
   7. $231 \div 5$
   8. $346 \div 2$

6. Suggest another situation as described in Problem #9. "The trucking company always wants the same full load in all trucks and calls ahead of time to let you know what size truck is being sent. Your job is to figure out how many trucks are needed to ship the whole load. For example, if each truck holds cartons and only 12, how many trucks would be needed to haul 304 cartons?" For many students it becomes tedious to fair share cartons into 12 sets, so another way to think about the problem is to ask how many sets of 12 can be separated from 304? This suggests the commutative property of multiplication when you look at the relationship between the divisor and the quotient.

   9. $304 \div 12$
  10. $643 \div 128$

Discussion

1. Describe how you solved problem #1 using manipulatives. [Cartons were fair shared into 3 trucks.]
2. Describe how you solved Problem #9. [Three hundred and four cartons were grouped in sets of 12.]
3. Compare how the two problems are similar and how they are different. [One asks how many cartons per truck; the other asks how many trucks for all the cartons. Both require the division process because a large number is separated into smaller equal sets. In the process of division, two questions can be posed. How many are in a set? and How many sets?]
4. How are the procedures alike or different for each problem? [See *Employee Loading Instructions.*]

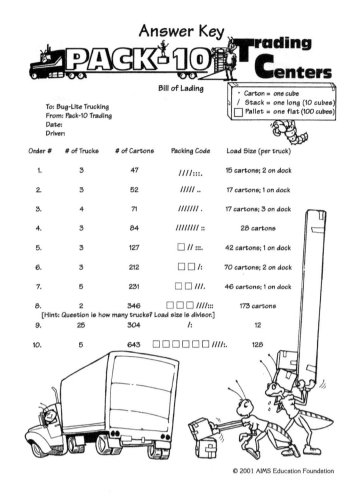

Answer Key

PACK-10 Trading Centers

Bill of Lading

- Carton = one cube
- / Stack = one long (10 cubes)
- ☐ Pallet = one flat (100 cubes)

To: Bug-Lite Trucking
From: Pack-10 Trading
Date:
Driver:

Order #	# of Trucks	# of Cartons	Packing Code	Load Size (per truck)
1.	3	47	////:::.	15 cartons; 2 on dock
2.	3	52	///// ..	17 cartons; 1 on dock
3.	4	71	//////// .	17 cartons; 3 on dock
4.	3	84	//////// ::	28 cartons
5.	3	127	☐ // :::.	42 cartons; 1 on dock
6.	3	212	☐☐ /:	70 cartons; 2 on dock
7.	5	231	☐☐ ///.	46 cartons; 1 on dock
8.	2	346	☐☐☐ ////:::	173 cartons
9.	25	304	/:	12
10.	5	643	☐☐☐☐☐ ////:.	128

[Hint: Question is how many trucks? Load size is divisor.]

* Reprinted with permission from *Principles and Standards for School Mathematics*, 2000 by the National Council of Teachers of Mathematics. All rights reserved.

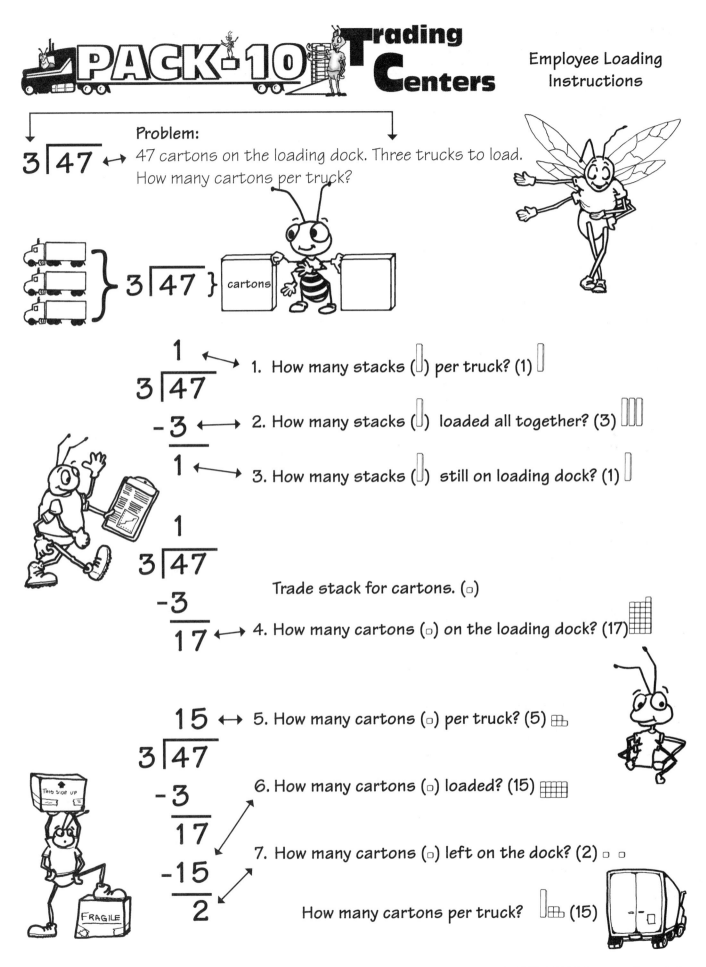

Problem:

3)47 ↔ 47 cartons on the loading dock. Three trucks to load. How many cartons per truck?

cartons

1. How many stacks (▯) per truck? (1) ▯

2. How many stacks (▯) loaded all together? (3) ▯▯▯

3. How many stacks (▯) still on loading dock? (1) ▯

Trade stack for cartons. (▫)

4. How many cartons (▫) on the loading dock? (17)

5. How many cartons (▫) per truck? (5)

6. How many cartons (▫) loaded? (15)

7. How many cartons (▫) left on the dock? (2) ▫ ▫

How many cartons per truck? (15)

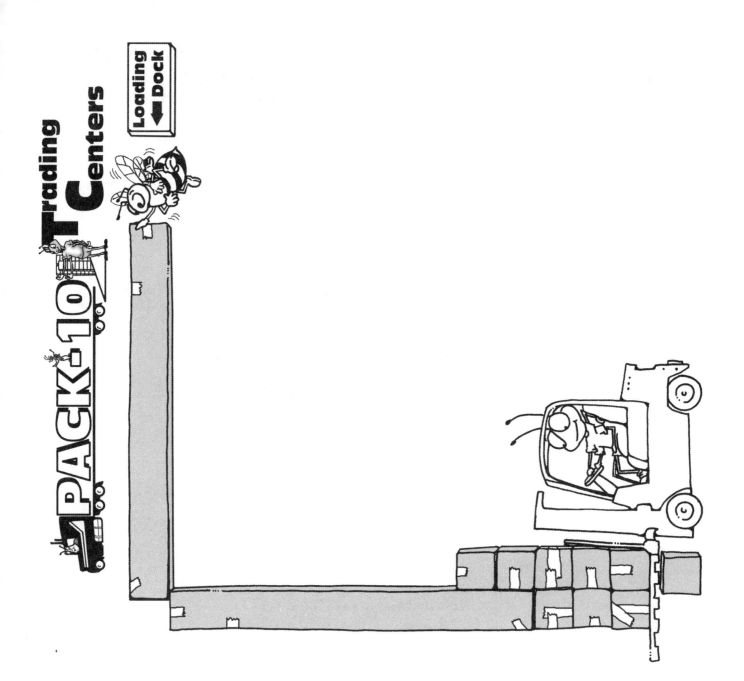

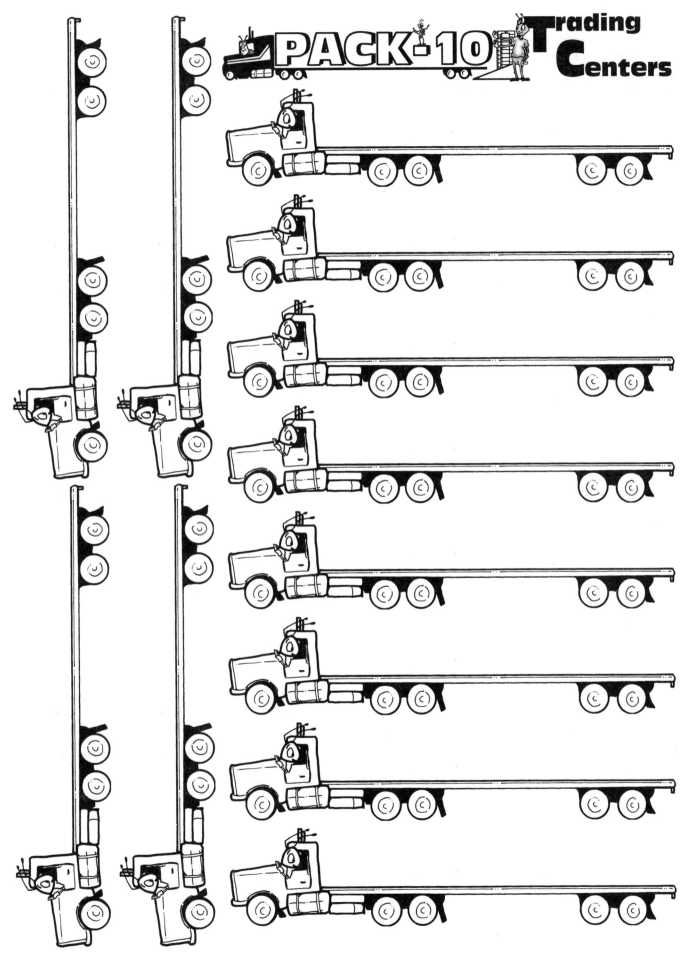

PACK·10 Trading Centers

Bill of Lading

> • Carton = one cube
> / Stack = one long (10 cubes)
> ▢ Pallet = one flat (100 cubes)

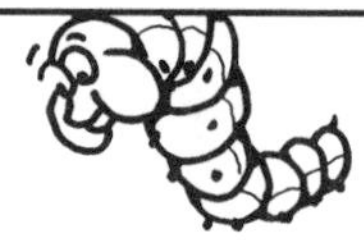

To: Bug-Lite Trucking
From: Pack-10 Trading
Date:
Driver:

Order #	# of Trucks	# of Cartons	Packing Code	Load Size (per truck)
1.	3	47	////:::.	15 cartons; 2 on dock
2.				
3.				
4.				
5.				
6.				
7.				
8.				
9.				
10.				

Clearing the Table

Topic
Basic multiplication facts
Patterns in multiples
Divisibility rules

Learning Goals
Students will:
- recognize patterns in the multiplication table, and
- use mathematical reasoning to explore divisibility rules.

Guiding Document
NCTM Standards 2000*
- *Use computation tools and strategies fluently and estimate appropriately*
- *Identify and use relationships between operations, such as division as the inverse of multiplication, to solve problems*

Materials
Multiplication Table for each student (through 12s)
Colored pencils
One large (wall-size), laminated multiplication table
Pad of small ($1\frac{1}{2}$" x 2") sticky notes

Background Information
In this activity we will limit our discussion of patterns and divisibility rules to facts through the tens. For 11s and 12s, see *Just the Answers*.

Management
Laminate a large multiplication chart and mount on a wall in the classroom. Have available sticky notes to "hide" or cover up each number as it becomes part of a pattern.

Procedure
1. Distribute one *Multiplication Table* to each student in a group of four.
2. Ask students to look for and mark with a colored pencil any patterns in rows and columns of numbers in the multiplication table. Encourage them to move beyond the obvious patterns such as the repeating 0,5 in the multiples of 5.

3. Where possible, ask students to connect recognized patterns to divisibility rules. For example, in the rows and columns of multiples of two, the numbers always end in an even number (0, 2, 4, 6, 8). Therefore, to determine if a number is divisible by two, it simply must end in an even number. The reference to "end" means in the units place.
4. Have students share their patterns and their processes of discovery.
5. On a classroom wall chart of the multiplication table, block out or cover with a sticky note, each series of facts for which there is a helpful set of rules for remembering the fact.
6. Challenger 6! Ask students to discover the divisibility rule for 6.
7. For practice, play *Division Dominoes*.

Discussion
1. Which columns and rows of numbers were easiest to identify? [1, 2, 5] Why? [Because number patterns repeated every other number such as 0, 5, 0, 5, etc., or numbers were easily recognized such as 1, 2, 4, 6, 8, repeat]
2. How do these easy patterns connect to divisibility? [All numbers that end in 0 or 5 are divisible by

five. And all numbers that end in an even number are divisible by two.]

consecutive numbers in the tens place and its inverse in the units place for the column of nines.

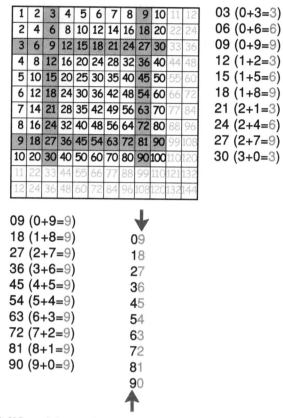

03 (0+3=3)
06 (0+6=6)
09 (0+9=9)
12 (1+2=3)
15 (1+5=6)
18 (1+8=9)
21 (2+1=3)
24 (2+4=6)
27 (2+7=9)
30 (3+0=3)

09 (0+9=9)   09
18 (1+8=9)   18
27 (2+7=9)   27
36 (3+6=9)   36
45 (4+5=9)   45
54 (5+4=9)   54
63 (6+3=9)   63
72 (7+2=9)   72
81 (8+1=9)   81
90 (9+0=9)   90

4. What did you discover about the Challenger 6 rule? [Responses could include that all numbers are even; thus divisible by two. And all digits add up to a number divisible by three—3, 6, 9. Therefore, a number divisible by 6 must be divisible by both 2 and 3.]

06 (0+6=6)
12 (1+2=3)
18 (1+8=9)
24 (2+4=6)
30 (3+0=3)
36 (3+6=9)
42 (4+2=6)
48 (4+8=12)
54 (5+4=9)
60 (6+0=6)

3. What do you notice about column three and nine? [Threes skip by three and the digits add up to a number divisible by 3—3, 6, 9. In the column of nines, the digits add to 9—also divisible by 3.] Students may also notice the pattern of

5. Which columns and rows of numbers remain?
   [4, 7, 8]

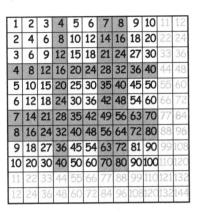

1	2	3	4	5	6	7	8	9	10	11	12
2	4	6	8	10	12	14	16	18	20	22	24
3	6	9	12	15	18	21	24	27	30	33	36
4	8	12	16	20	24	28	32	36	40	44	48
5	10	15	20	25	30	35	40	45	50	55	60
6	12	18	24	30	36	42	48	54	60	66	72
7	14	21	28	35	42	49	56	63	70	77	84
8	16	24	32	40	48	56	64	72	80	88	96
9	18	27	36	45	54	63	72	81	90	99	108
10	20	30	40	50	60	70	80	90	100	110	120
11	22	33	44	55	66	77	88	99	110	121	132
12	24	36	48	60	72	84	96	108	120	132	144

6. How could we find a helpful pattern for fours?
   [Any multiple of 4 is an even number that can be divided by 2—twice.]

7. Which three facts still remain uncleared from the table? [7 x 7, 7 x 8, 8 x 8]

8. What hints or clues would be helpful with these three facts? [Two are square numbers—49, 64. And 56 = 7 x 8 is four consecutive numbers!]

9. The column or row of eights is interesting. What patterns can you find? [In the units place, the digits repeat in a series 8, 6, 4, 2, 0; in the tens place the digits are consecutive and the last number in each set begins the next set followed by four consecutive numbers]

10. The column (row) of sevens is very difficult to find a pattern that is helpful.)

Multiplication Table

1	2	3	4	5	6	7	8	9	10	11	12
2	4	6	8	10	12	14	16	18	20	22	24
3	6	9	12	15	18	21	24	27	30	33	36
4	8	12	16	20	24	28	32	36	40	44	48
5	10	15	20	25	30	35	40	45	50	55	60
6	12	18	24	30	36	42	48	54	60	66	72
7	14	21	28	35	42	49	56	63	70	77	84
8	16	24	32	40	48	56	64	72	80	88	96
9	18	27	36	45	54	63	72	81	90	99	108
10	20	30	40	50	60	70	80	90	100	110	120
11	22	33	44	55	66	77	88	99	110	121	132
12	24	36	48	60	72	84	96	108	120	132	144

What patterns in the products of six help you invent a rule for divisibility?

Your Thinking: (doodle here)

X	6
1	6
2	12
3	18
4	24
5	30
6	36
7	42
8	48
9	54
10	60
11	66
12	72

Record your rule for numbers divided by six.

Test it here with three trials. Name three numbers between 100-500 that are evenly divided by six. Show that your rule works.

Topic
Prime and composite numbers

Learning Goals
Students will
- explore prime and composite numbers by constructing rectangles with square area tiles, and
- organize their findings and generalize that some numbers are divisible only by one and themselves (prime) and others have multiple divisors.

Guiding Document
NCTM Standards 2000*
- *Describe classes of numbers (e.g., odds, primes, squares, and multiples) according to characteristics such as the nature of their factors*
- *Recognize equivalent representations for the same number and generate them by decomposing and composing numbers*

Materials
For each student:
12–15 square area tiles
scissors
2-cm grid paper
index cards, 3" x 5"

Math
Number sense
    prime/composite

Integrated Processes
Observing
Comparing and contrasting
Generalizing

Background Information
Whole numbers can be represented pictorially in two-dimensional form by constructing rectangular arrays with square area tiles. As students build the rectangular arrays, they will discover that some numbers can only be made using lengths of that number and one. These are the prime numbers. A prime number is any number greater than one that has only two divisors, one and itself. Composite numbers are numbers that have more than two divisors. There will be more than one array for each composite number.

The opportunity for the development of mathematical language is a powerful piece of this experience. Such terms as factor, multiple, divisor, prime, composite, compose, decompose, prime factors, and prime factorization are integral to this lesson.

Management
1. Students should work in pairs or teams of four to collect and share data.
2. You may wish to adjust the range of whole numbers that students search. For third and fourth graders, 2–12 is suggested. Perhaps for fifth and sixth graders, the search should be expanded to 20. This would include the most frequently used prime factors—2, 3, 5, 7, 11, 13, 17, and 19.

Procedure
1. Distribute tiles to students and explain that they are to build as many different rectangles as are possible for each whole number between 2 and 12.
2. Demonstrate the rectangles possible for 4. Then explain that a line of 4 tiles and a square region of four tiles are the only two possible rectangles that can be constructed from four tiles.

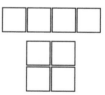

3. As rectangles are completed, ask students to cut a matching rectangle from 2-cm square grid paper.
4. Direct them to match paper rectangles with each number between 2 and 12 representing the total number of square units or tiles in each rectangle.
5. Ask students to make an organized, symbolic/numeric record of each rectangle by drawing a picture of the array and recording the length and width of each array. Demonstrate that rectangles for 4 can be written 1 x 4 and 2 x 2.
6. Ask students to record observations and look for generalizations about the rectangles and the numbers they represent.
7. Help students confirm the idea that a prime number is any number greater than one with factors (divisors) of only one and itself. Ask them to

verify the prime numbers between 2 and 12 for which they built rectangles. Each of these primes has only one rectangle possible. The remaining numbers have two or more possible rectangles with factors (divisors) other than one and itself and represent composite numbers.

Discussion

1. How are the rectangles that you constructed alike and/or different from each other? [All rectangles have four sides. Some are squares. Some are only one unit wide while others are two or more units wide.]

2. What role do the sides of the rectangle play? [They determine the factors or divisors.]

3. Use the cut paper rectangles to compare the picture of each rectangle to its symbolic record. How are the factors related to the rectangular array? [The factors or divisors represent the two dimensions of the rectangle.]

4. When you match all of the rectangles to the numbers between 2 and 12, what observations can be made about the relationships between the numbers and the pictures? [The fewer the number of rectangles, the fewer the number of factors or divisors of that number. Some numbers have only two divisors, 1 and itself. Others have several more divisors.]

5. Challenge students to examine the *100 Chart* and determine which numbers are prime between 1 and 100. Cross out numbers in the chart that are divisible by a factor other than 1 or itself. What are the only four divisors that need to be considered to eliminate all the composite numbers in that chart? [2, 3, 5, 7.] What is the common characteristic of these four numbers? [They are the first four prime numbers.]

6. Provide practice factoring numbers into their prime factors by doing *Factor—Ease*. (See activity page). Explain that a number can be decomposed or broken down into its prime factors through a series of steps.

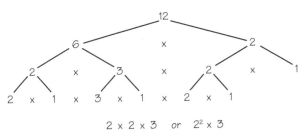

$$2 \times 2 \times 3 \quad \text{or} \quad 2^2 \times 3$$

Have students try other numbers such as 12, 42, 39, etc. Distribute index cards and assign a number to be factored to each card. Compare and organize the cards. Look for patterns. Make a generalization about the factors of all the numbers less than 100.

Extensions

1. Have students play around with the idea that any even number greater than 2 can be expressed as the sum of two prime numbers. This is known as Goldbach's Conjecture and has never been proven or disproven. Goldbach lived in the early 16th century so lots of time has passed for young mathematicians to work on this problem! Make a list of all the primes between 1 and 100. What patterns or "conjectures" can you invent based on relationships you observed?

2. Challenge students to explore the sums of the divisors of any number in search of the "perfect" number. A perfect number is a number whose divisors smaller than itself add up to the number itself. For example, 6 is a perfect number because all its divisors smaller than six, 1, 2, 3, add up to 6. Six is the first perfect number. Can students find the next perfect number? (Hint: It is less than 100.) This idea is cleverly presented in *Math For Smarty Pants* by Marilyn Burns published by Little, Brown & Co., 1982. ISBN 0-316-11739-0. The article is called, "Some Numbers are More Perfect Than Others," p. 124 – 126.

One Number INDIVISIBLE

Number	Picture	Symbol		Prime/Composite
2		1 x 2		Prime
3		1 x 3		Prime
4		1 x 4 2 x 2		Composite
5		1 x 5		Prime
6		1 x 6 2 x 3	6 x 1 3 x 2	Composite
7		1 x 7		Prime
8		1 x 8 2 x 4	8 x 1 4 x 2	Composite
9		1 x 9 3 x 3	9 x 1	Composite
10		1 x 10 2 x 5	10 x 1 5 x 2	Composite
11		1 x 11		Prime
12		1 x 12 2 x 6 3 x 4	12 x 1 6 x 2 4 x 3	Composite

AIMS JULY/AUGUST

* Reprinted with permission from *Principles and Standards for School Mathematics,* 2000 by the National Council of Teachers of Mathematics. All rights reserved.

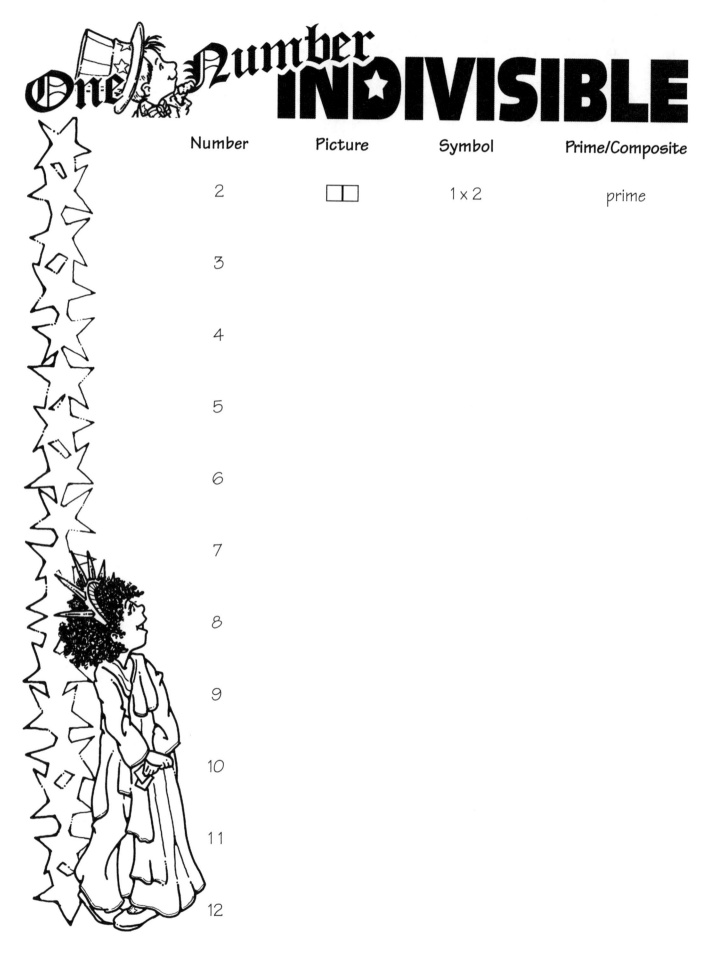

One Number INDIVISIBLE

Number	Picture	Symbol	Prime/Composite
2	□□	1 x 2	prime
3			
4			
5			
6			
7			
8			
9			
10			
11			
12			

Select eight composite numbers from the Hundreds Chart and picture their separation into factors until all factors are prime. Use the back of the paper if necessary.

```
            6
          /   \
        2   x   3
       / \     / \
      1 x 2 x 3 x 1
```

Why do you think some numbers have only two levels of separation and others have more?

How could you use exponents to abbreviate or consolidate your notation?

1	2	3	4	5	6	7	8	9	10
11	12	13	14	15	16	17	18	19	20
21	22	23	24	25	26	27	28	29	30
31	32	33	34	35	36	37	38	39	40
41	42	43	44	45	46	47	48	49	50
51	52	53	54	55	56	57	58	59	60
61	62	63	64	65	66	67	68	69	70
71	72	73	74	75	76	77	78	79	80
81	82	83	84	85	86	87	88	89	90
91	92	93	94	95	96	97	98	99	100

Cross out all numbers that are not prime.

Playful, Intelligent Practice
Multiplication/Division

Elementary classroom teachers understand the importance of providing practice of basic skills, specifically those that deal with basic operations on whole numbers. While the adage *practice makes perfect* makes sense to us, over the last 15 years there is considerable support for the idea that "drilling" for periods longer than 10 minutes a day may be counter-productive. Let's think about how practice could be both playful and intelligent at the same time, therefore maximizing the 10 minutes of focused practice!

In conversations with teachers, we've learned that certain characteristics of "drill and practice" tasks are desirable. Such features include: an element of "playfulness," minimum preparation on the part of the teacher, time efficient, context of mental stimulation and exercise, self-correcting, and interesting and motivating. Other byproducts such as pressure under timed tests and strong negative feelings about mathematics and self are not desirable.

Who Has?, Crazy Clues, and *Division Dominoes* are mentally stimulating because of the "playful" context in which they are presented. A whole class experience may be facilitated by a teacher, a student, or a classroom assistant. Students may participate individually or in pairs. *Who Has?* and *Crazy Clues* may be expanded by teachers and students by "writing their own."

Who Has?

Directions:

Distribute one card to each student, or several cards to pairs of students. Select a student to "begin" by reading his/her card aloud — "I have 33. Who has that divided by 3?" The person holding the answer to the card responds by reading aloud the card with the appropriate response. The game is self-generating and ends when the cycle returns to the beginning card; in this case, 33.

Designing your own game:

Determine the number of cards/students in a complete set.

Each card generates a number (an answer) that may be used only once. The last card in the set returns the game to the number on the beginning card.

Keep in mind the range of numbers and operations that are grade level appropriate. One could begin to construct a game for 30 students simply by randomly listing 30 counting numbers and then "connecting" them by ascribing an operation to produce the next number. Some students like to combine two or more operations to make the problem more difficult. It is far more interesting if the answers are "out of order" so that students are unaware of the next right answer. It is also worthwhile to examine the kinds of questions that students need to ask when generating their own game. What kinds of thinking do students engage in when considering which operations are appropriate to get from 60 to 45 for example?

Card Set

I have 33. Who has that divided by 3?
I have 11. Who has that plus 9?
I have 20. Who has that times 3?
I have 60. Who has that minus 15?
I have 45. Who has that divided by 5?
I have 9. Who has that times 9?
I have 81. Who has that plus 9?
I have 90. Who has that minus 20?
I have 70. Who has that minus 6?
I have 64. Who has that minus 16?
I have 48. Who has that divided by 2?
I have 24. Who has that divided by 4?
I have 6. Who has that plus 10?
I have 16. Who has that minus 1?
I have 15. Who has that plus 6?
I have 21. Who has that times 3?
I have 63. Who has that divided by 9?
I have 7. Who has that plus 3?
I have 10. Who has that minus 7?
I have 3. Who has that times 6?
I have 18. Who has that plus 12?
I have 30. Who has that divided by 6?
I have 5. Who has that times 8?
I have 40. Who has that minus 4?

I have 36. Who has that divided by 9?
I have 4. Who has that plus 9?
I have 13. Who has that plus 12?
I have 25. Who has that times 3?
I have 75. Who has that minus 3?
I have 72. Who has that divided by 9?
I have 8. Who has that plus 9?
I have 17. Who has that plus 16?

Card Set

I have 11. Who has that times 5; plus one more?

I have 56. Who has that divided by 8?

I have 7. Who has that squared; plus 3?

I have 52. Who has that take away 12; then divided by 4?

I have 10. Who has that squared; then divided by 2?

I have 50. Who has that plus 13; divided by 7?

I have 9. Who has the product of that times 8?

I have 72. Who has that take away 13?

I have 59. Who has that divided by 8?

I have 7 remainder 3. Who has 7 times 4 plus 3 more?

I have 31. Who has that take away 4; and then divided by 9?

I have 3. Who has that times 12; plus 8 more?

I have 44. The divisor is 11. Who has the quotient?

I have 4. Who has that multiplied by 6 and again by 2?

I have 48. Who has that divided by 8?

I have 6. Who has that squared; plus 7 more?

I have 43. Who has that divided by 8?

I have 5 remainder 3. Who has the product of 7 and 5; increased by 3 more?

I have 38. Who has that divided by 2; then decreased by 4?

I have 15. Who has that times 2; take away 1?

I have 29. Who has that take away 14; then divided by 3?

I have 5. Who has that squared; then doubled; plus 12 more?

I have 62. Who has that plus 2; divided by 8?

I have 8. Who has that times 2; plus 7 more?

I have 23. Who has that tripled; then take away 2?

I have 67. Who has that plus 5; then divided by 6?

I have 12. Who has that doubled; plus 2?

I have 26. Who has that plus 8; divided by 2?

I have 17. Who has that times 2; plus 1?

I have 35. Who has that take away 3; divided by 8; and multiplied by 6?

I have 24. Who has that divided by 4; then squared?

I have 36. Who has that divided by 2?

I have 18. Who has that plus 5 sets of 10?

I have 68. Who has that take away 4; then divided by 4?

I have 16. Who has that times 3 times 2?

I have 96. Who has that take away 8; divided by 8?

(Repeat from the beginning).

I have 52.

Who has that take away 12; then divided by 4?

I have 72.

Who has that take away 13?

I have 7.

Who has that squared; plus 3?

I have 9.

Who has the product of that times 8?

I have 56.

Who has that divided by 8?

I have 50.

Who has that plus 13; divided by 7?

I have 11.

Who has that times 5; plus one more?

I have 10.

Who has that squared; then divided by 2?

I have 3.

Who has that times 12; plus 8 more?

I have 6.

Who has that squared; plus 7 more?

I have 31.

Who has that take away 4; and then divided by 9?

I have 48.

Who has that divided by 8?

I have 7 remainder 3.

Who has 7 times 4 plus 3 more?

I have 4.

Who has that multiplied by 6 and again by 2?

I have 59.

Who has that divided by 8?

I have 44.

The divisor is 11. Who has the quotient?

I have 15.

Who has that times 2; take away 1?

I have 8.

Who has that times 2; plus 7 more?

I have 38.

Who has that divided by 2; then decreased by 4?

I have 62.

Who has that plus 2; divided by 8?

I have 5 remainder 3.

Who has the product of 7 and 5; increased by 3 more?

I have 5.

Who has that squared; then doubled; plus 12 more?

I have 43.

Who has that divided by 8?

I have 29.

Who has that take away 14; then divided by 3?

I have 26.

Who has that plus 8; divided by 2?

I have 36.

Who has that divided by 2?

I have 12.

Who has that doubled; plus 2?

I have 24.

Who has that divided by 4; then squared?

I have 67.

Who has that plus 5; then divided by 6?

I have 35.

Who has that take away 3; divided by 8; and multiplied by 6?

I have 23.

Who has that tripled; then take away 2?

I have 17.

Who has that times 2; plus 1?

I have 96.

Who has that take away 8; divided by 8?

I have 16.

Who has that times 3 times 2?

I have 68.

Who has that take away 4; then divided by 4?

I have 18.

Who has that plus 5 sets of 10?

Crazy Clues

??? ? ?

Directions:

Select a person such as the teacher or a cross-age tutor to read the "problem" aloud. Problems are challenging because they incorporate interesting ideas and contexts. For instance, "The number of legs on six elephants and a penguin" requires that students think about the number of legs on an elephant and also on a penguin and then be able to multiply and combine the answers.

"I'll take a silent, raised hand" helps manage the enthusiastic and eager responses of students. Call on a student to give only the answer. If correct, move on to next question. If incorrrect, call on another student. Because of the playful nature of the problems, students forget that they are doing mental math and gaining practice on basic facts and operations. It is also interesting to note that in addition to the "practice" feature, students are also learning to sort and use relevant information and becoming skilled at analyzing and selecting relevant information and prioritizing sequence of operations — all in their heads!

Designing your own game:

Students and teachers alike can be very creative in the design of *Crazy Clues*. One could "set the stage" for this by listing, with students, familiar objects and events and a feature that is "countable." For instance, ears on a rabbit, legs on a table, and so forth. Then add interesting information to raise both the complexity and the interest in the situation. The number of ears on 6 white rabbits with fluffy tails. Students also like to increase the number of objects or events being collected (mentally); perhaps begin with two and then add a third and then a fourth. Excellent potential exists for connections to children's literature, science and social studies trivia, and language arts elements.

Introductory Set, Crazy Clues

The number of legs on six elephants and a penguin. (26)

The number of letters in STOP AND GO. (9)

The number of wheels on four cars, two tricycles, and a little red wagon. (26)

The number of minutes in an hour minus the number of days in September. (30)

The number of hangers needed for three shirts, four pants, and a pair of socks. (7)

The number of eyes on three needles, a Cyclops, and Mississippi. (8)

The number of ears on six rabbits and a half-dozen ears of corn. (18)

The number of blind mice and months in a year. (15)

The number of sunrises and sunsets in a week. (14)

The number of Snow White's dwarfs and Santa's reindeer. (15)

The number of little pigs that went to market plus the number of Musketeers. (4)

The number of Ali Baba's thieves and Cinderella's stepsisters. (42)

The number of fingers on two boys and a little girl with blonde hair. (30)

The number of players in two football games. (44)

The number of eyes on eight owls minus the noses on two skunks. (14)

Crazy Clues You Can Count On

Beginning Facts Set

The number of wheels on three wagons times the wings on four planes.
[12 x 8 = 96]

The number of pennies in a dime times the days in January.
[10 x 31 = 310]

The number of keys on a piano divided by the sides of an octagon.
[88 ÷ 8 = 11]

The number of players on a football team times the teams in four games. [11 x 8 = 88]

The number of sunsets in a week times the months in a year. [7 x 12 = 84]

The number of squares on a checkerboard divided by the sides on two squares. [64 ÷ 8 = 8]

The number of Back Street Boys times the notes on a scale. [5 x 8 = 40]

The number in half a dozen, squared. [6 x 6 or 6^2 = 36]

The number of bases on a diamond times the players on a team.
[4 x 9 = 36]

The number of toes on a baby times the bills on 4 ducks. [10 x 4 = 40]

More Difficult—Double Digit Operations

The number of ears on eight rabbits times the eyes on two skunks. [16 x 4 = 64]

The number of cards in a deck divided by a baker's dozen. [52 ÷ 13 = 4]

The number of ounces in a pound times the singers in a trio. [16 x 3 = 48]

The number of weeks in a year divided by the quarters in a dollar. [52 ÷ 4 = 13]

The number of holes on a golf course times the feet in a yard. [18 x 3 = 54]

The number of degrees (Fahrenheit) at which water freezes times the mittens on three Eskimos. [32 x 6 = 192]

The number of degrees in a circle divided by eggs in a dozen. [360 ÷ 12 = 30]

The number of quarts in two gallons times the number of letters in GOT MILK. [8 x 7 = 56]

The number of minutes in three-fourths of an hour times the number of digits in your zip code. [45 x 5 = 225]

The number of digits in your area code times the number of zeros in one hundred, times the stripes in the American flag. [3 x 2 x 13 = 78]

Division Domino games possess these desirable features.

1. The games can be played solitaire or by two or more players in competition. The solitaire format has several advantages. Losing is less painful since it is a private experience. Furthermore, there is something about solitaire games that attracts certain people to play time and again.

2. The basic games are self-correcting, transferring the workload for checking the results from the teacher to the student.

3. When an error occurs, a system exists for correcting the error and completing the game successfully.

4. The chance of winning, and hence the frustration level, can be adjusted to fit the needs of individual students making it easier for slow students and more challenging for better students.

5. The focus can be on the quotient disregarding the remainder or on the remainder disregarding the quotient.

6. *Division Dominoes* emphasize mental computation and scanning while reviewing basic division facts. Scanning, in which many quotients or remainders are computed in search of a desired result, is rarely practiced in traditional curricula even though it is a very practical skill in real-life situations.

7. In any game, players will make many more computations than those which are finally used to create pairs. They will also develop strategies for "holding" information in their minds while making other computations. A trial run by the teacher will quickly surface this feature.

Quotient Solitaire

Shuffle the dominoes thoroughly. Deal out the entire set face up in 10 stacks. After the first 10 have been placed down, continue to cover them in order. Since there are 54 dominoes, six of the stacks will have five dominoes and four will have six at the beginning of play. This arrangement is the most convenient for viewing all 10 dominoes.

Remove dominoes **in pairs** and **only in pairs. In each pair the quotient disregarding the remainder must match.** For example, the domino 7⟌39 will match 9⟌52 since the quotient disregarding the remainder is 5 in both cases.

As pairs are removed, they should be criss-crossed on a pile so that they can easily be identified as pairs later to locate an error.

Continue to remove pairs. As soon as one pile is completely removed, break any other pile at any point and move some dominoes into the vacated position in order to keep 10 dominoes showing at all times until the very last stages of the game when fewer than 10 remain in play.

This set forms 27 pairs, sometimes in different combinations.

If all of the dominoes pair out, the computations have been made correctly and the player is the winner. If at any time during the game no pairs are showing (be sure to check and double check), the player has lost against the deck.

If the last two, four, or six dominoes, etc. do not pair out, then the player has made one or more errors. To check, examine all of the pairs that have been removed (the reason pairs were criss-crossed). When a mismatched pair is located, return the dominoes to play. If only one error were made, the dominoes will now pair out. If more than one error were made, this procedure must be continued until the dominoes pair out.

To increase the chance of winning, increase the number of stacks. To make winning more challenging, decrease the number of stacks to nine, then to eight, etc. Gifted students and students who have thoroughly mastered these basic facts will want to accept this challenge. The odds of winning increase as the number of stacks is increased, and the odds of winning decrease if the number of stacks is decreased.

Remainder Solitaire

The rules are the same as for Quotient Solitaire except that students will match remainder, disregarding the whole number in the quotient. The dominoes pair out as in Quotient Solitaire. This game is self-correcting.

Caution: Be sure that no dominoes are lost and each set is complete. If incomplete, the self-correcting feature is destroyed.

Competitive Division Dominoes

Quotient Domino War

Two to four players

Deal out all the dominoes to the players. Each player places his/her dominoes face down in a stack.

During each turn, each player turns the top domino face up. The player with the largest quotient disregarding the remainder is the winner. If it is a tie, those dominoes are placed in a discard pile. If a player has a quotient disregarding remainder that is five greater than those of any opponent, that player may claim all the cards then in the discard pile.

The game ends when all dominoes have been played.

The player with the largest number of dominoes at the end of the game is the winner.

This variation is not self-correcting.

Quotient Pairs

Two to four players

Deal five dominoes to each player. Place three dominoes face up on the table. Place the remaining dominoes face down on the "bone pile."

The first player inspects the dominoes on the table and those in hand to find a pair with matching quotients disregarding the remainder. The pair may consist of any two, whether both are from the hand, the table, or one of each. If such a pair is found, it is claimed by the player. This will decrease the number of dominoes on the table if one or two dominoes are paired from those on the table. If the player finds no pair, the player must place one domino from the hand onto the table (thus increasing the number on the table), this ending the turn. Only one pair may be claimed in a given turn.

At the end of each turn, the player picks up one or two dominoes from the bone pile so that five dominoes are again in hand. Only in the final stages of the game when the bone pile is depleted will fewer than five dominoes be in hand.

Should all of the dominoes be removed from the table at some stage of the game, the next player must lay down a domino and that ends the turn.

Each pair claimed by a player scores one point. The player with the largest number of claimed pairs is the winner.

Players should explain their solutions to each other so that all can check to make sure the pairing has been made correctly. If a player makes an error, the turn ends and the player must place one domino on the table.

The game is self-correcting. If all pairs have been formed correctly, all of the dominoes will pair out.

Remainder Domino War

The rules are the same as those for Quotient Domino War except that students will match remainders, disregarding the whole number in the quotient. This format is not self-correcting.

Remainder Pairs

The rules are the same as for Quotient Pairs except that students will match remainders, disregarding the whole number in the quotient. The dominoes pair out as in Quotient Pairs. This game is self-correcting.

Caution: Be sure that no dominoes are lost and each set is complete. If incomplete, the self-correcting feature is destroyed.

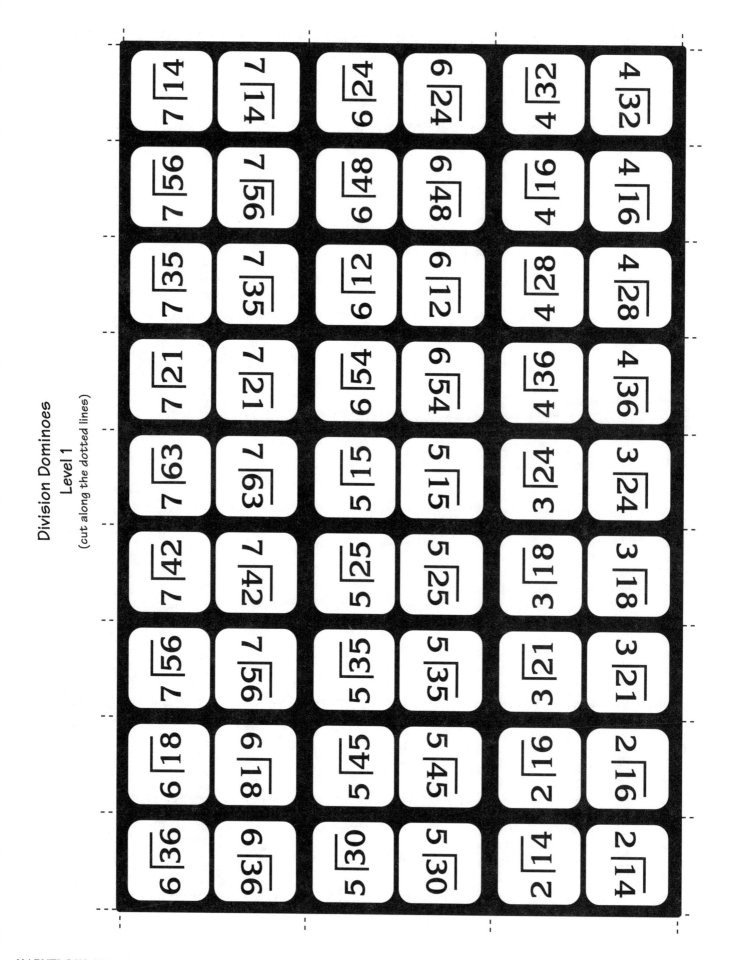

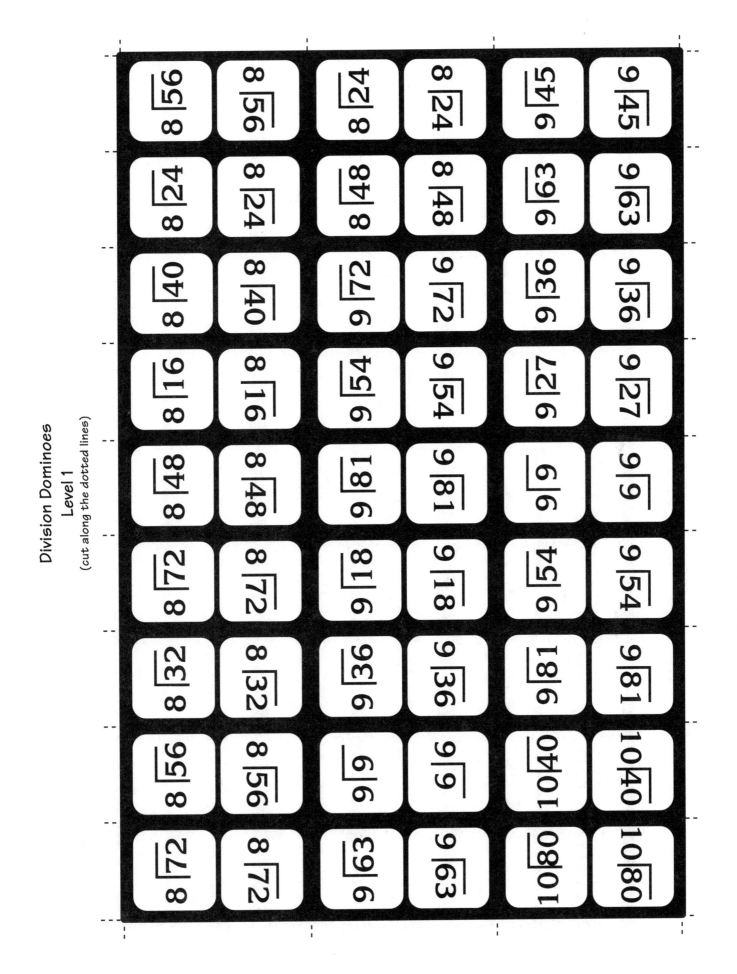

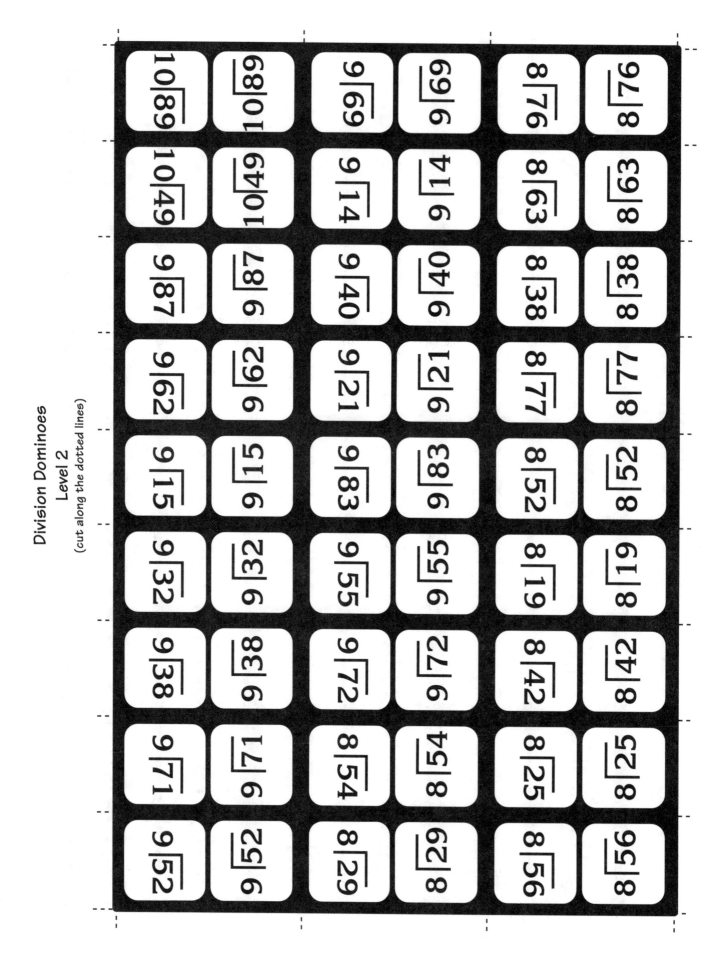

Division Dominoes
Level 2
(cut along the dotted lines)

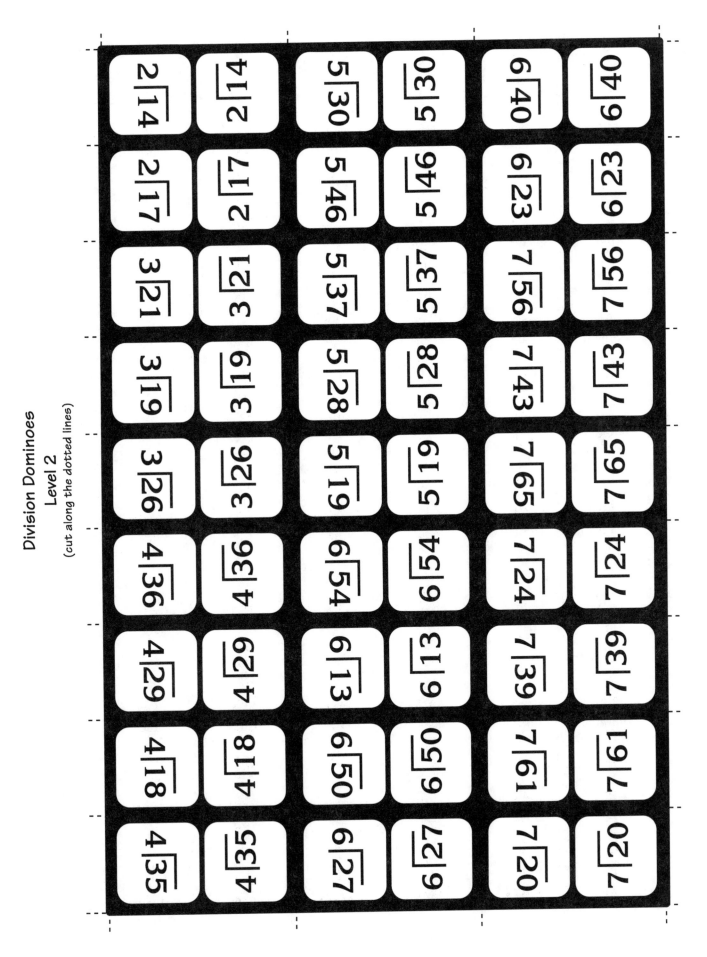

Division Dominoes
Level 2
(cut along the dotted lines)

2⟌14	2⟌14
2⟌17	2⟌17
3⟌21	3⟌21
3⟌19	3⟌19
3⟌26	3⟌26
4⟌36	4⟌36
4⟌29	4⟌29
4⟌18	4⟌18
4⟌35	4⟌35
5⟌30	5⟌30
5⟌46	5⟌46
5⟌37	5⟌37
5⟌28	5⟌28
5⟌19	5⟌19
6⟌54	6⟌54
6⟌13	6⟌13
6⟌50	6⟌50
6⟌27	6⟌27
6⟌40	6⟌40
6⟌23	6⟌23
7⟌56	7⟌56
7⟌43	7⟌43
7⟌65	7⟌65
7⟌24	7⟌24
7⟌39	7⟌39
7⟌61	7⟌61
7⟌20	7⟌20

Topics
Problem solving, Estimation

Learning Goals
Students will:
- devise a strategy for calculating a close approximation of the number of objects in a container without counting each one, and
- communicate mathematical ideas by organizing and creating a visual display of their solutions and problem-solving strategies.

Guiding Document
NCTM Standards 2000*
- *Develop fluency in adding, subtracting, multiplying, and dividing whole numbers*
- *Develop and use strategies to estimate the results of whole-number computations and to judge the reasonableness of such results*
- *Apply and adapt a variety of appropriate strategies to solve problems*
- *Analyze and evaluate mathematical thinking and strategies of others*
- *Use the language of mathematics to express mathematical ideas precisely*
- *Create and use representations to organize, record, and communicate mathematical ideas*

Materials
8-10 resealable object bags (See *Management 1*)
Measuring tools (see *Management 4*)
Chart paper
Colored marking pens

Background Information
This activity is designed to strengthen students' ability to communicate effectively (through words, pictures, and symbols) things which have been

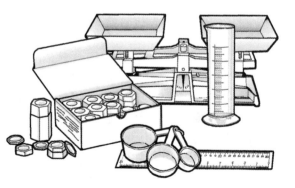

experienced in the real world in a problem-solving context. The "doing" process is relatively easy, matching an appropriate explanation to the process is difficult, and connecting actions to the arithmetic operations is the most difficult, but very powerful.

Teamwork is a crucial component in this activity. Selecting the process and solving the problem should be a shared experience. Therefore, group work and presentation of results on large chart paper is recommended. The learning takes place in a group setting and the communication of the experience provides a measure of what has been learned.

Management
1. You will need eight to 10 resealable "object bags" filled with a large number of similar small objects. Items could include dried beans, unpopped popcorn, small buttons, dried macaroni, etc. Fill the bags with numbers of objects appropriate for the grade level being taught. Suggested number ranges may include: 200-700 for grades 2 and 3, 700-1500 for grades 4-5, and 900-2500 for grades 6-7.
2. Pre-count all objects and keep a record of counted objects safely hidden.
3. Begin with smaller numbers rather than larger numbers to insure success. Keep objects uniform in size and shape at first. To increase the sophistication of the activity, increase the number and vary the shapes and sizes in one bag.
4. You will need to provide a variety of measuring tools for the students to use as they try to determine the number of objects in the bags. Tools could include balances, stacking masses, small measuring scoops, graduated cylinders, rulers, etc.

Procedure
1. Distribute a bag of objects to each group of four to five students. Have students observe and describe the contents of the bag. Have them list as many attributes as possible, such as color, shape, size, uniformity. Discuss which ones could affect students' judgments in guessing how many are in the bag.
2. Have each student make a wild guess as to the number of objects in the bag. Within the groups, have students find their average wild guesses which will be shared with the class later.
3. Instruct groups to brainstorm a strategy to determine more accurately the number of objects

in the bag without counting each one. Possible strategies such as counting or measuring the volume or mass of a part of the total number of objects should be discussed.

4. Hand out the chart paper and colored pens to each group. Have them describe, step by step, the procedures they decided to use and record them on the chart paper. Instruct groups to include their thinking and explanations of the strategies using pictures, words and symbols. Tell them to be sure to include the math!

5. As a class, compare the average wild guess (A.W.G.), a better guess resulting from strategic planning (B.G.), and the actual count (A.C.) for each group.

6. Evaluate the pluses and minuses of using a particular strategy for estimating numbers. Determine which group's method appeared to be the most effective and discuss possible reasons for this.

7. Have groups explain which arithmetic operations were useful and how.

Discussion

1. How do the sizes or shape of the objects affect the number of objects in a bag? [The larger and more irregular the objects, the fewer will fit in a bag.]

2. What was the range of guesses in each group, and how close was the average wild guess to the actual count or to the strategic guess?

3. Describe the strategy your group selected to determine the actual number of objects in the bag. [Many strategies are possible. See *Sample Strategy*.]

4. How did your group's method compare to the methods used by other groups?

5. Which group had the most effective strategy? How do you know? [Their better guess and their actual count were the closest.] Why do you think this is?

6. Which arithmetic operations were useful, and why?

7. What would happen to your strategy if the size or the shape of the objects were not uniform? How would you adjust?

8. Discuss amount of error. [Amount of error is expressed as the relationship between the better guess and the actual total count. This relationship can be expressed as a fraction, decimal, or percent.]

Sample Strategy

One potential strategy is cited here as an example of a very simple and powerful way to determine a better guess.

Description in Words	Math Symbols
• Use the balance to divide the whole set of beans into two equal parts.	B = all the Beans
• Each part is one half.	$B \div 2 = \frac{1}{2}$ Beans

• Return one half of the beans to the bag and use the balance to divide the remaining one half into two equal parts.

$$\frac{1}{2}B \div 2 = \frac{1}{4}B$$

• Return one part (one fourth) to the bag and again use the balance to divide the one fourth into two equal parts now called one eighth.

$$\frac{1}{4} \div 2 = \frac{1}{8}B$$

• Return one eighth of the beans to the bag and use the balance to divide the remaining one eighth beans into two equal parts.

$$\frac{1}{8} \div 2 = \frac{1}{16}B$$

• Place one sixteenth into the bag and count the remaining one sixteenth beans. Multiply the number in one part (130) by the number of equal parts (16) to get the total number in the bag (2080). Compare to the actual count (2085).

16 x 130 = 2080 Beans

This experience actually occurred as described in a classroom of reluctant math problem solvers. The feeling of successful reward was spontaneous and powerful when the discovery of the actual count proved them to be "off" by only 5 beans out of 2085.

* Reprinted with permission from *Principles and Standards for School Mathematics*, 2000 by the National Council of Teachers of Mathematics. All rights reserved.

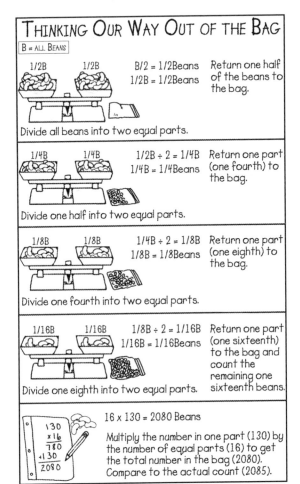

THINKING OUR WAY OUT OF THE BAG
B = ALL BEANS

1/2B 1/2B   B/2 = 1/2Beans   Return one half
              1/2B = 1/2Beans of the beans to the bag.
Divide all beans into two equal parts.

1/4B 1/4B   1/2B ÷ 2 = 1/4B   Return one part
              1/4B = 1/4Beans (one fourth) to the bag.
Divide one half into two equal parts.

1/8B 1/8B   1/4B ÷ 2 = 1/8B   Return one part
              1/8B = 1/8Beans (one eighth) to the bag.
Divide one fourth into two equal parts.

1/16B 1/16B 1/8B ÷ 2 = 1/16B Return one part
               1/16B = 1/16Beans (one sixteenth) to the bag and count the remaining one sixteenth beans.
Divide one eighth into two equal parts.

130
x16
780
+130
2080

16 x 130 = 2080 Beans
Multiply the number in one part (130) by the number of equal parts (16) to get the total number in the bag (2080). Compare to the actual count (2085).

THINKING OUR WAY OUT OF THE BAG

Class Results

Group	Objects	A.W.G. average wild guess	B.G. better guess	A.C. actual count
A				
B				
C				
D				
E				
F				
G				
H				
J				
K				

Thinking out of the bag!

1. Analyze the strategy of the group that had the closest results. What do you think made their planning more accurate than the others?

2. How would you alter the strategy for determining the count in your bag?

Topic
Problem solving with whole number operations

Learning Goals
Students will:
- apply thinking and problem-solving strategies to a real-world context, and
- apply whole number operations in complex, non-routine situations.

Guiding Document
NCTM Standards 2000*
- *Solve problems that arise in mathematics and other contexts*
- *Apply and adapt a variety of appropriate strategies to solve problems*
- *Compute fluently and make reasonable estimates*

Materials
Variety of printed resources for trip planning such as:
    state maps
    phone books
    travel information
    on-line Internet access

Background Information
This experience is designed to be a very open-ended problem-solving situation in which students have some choice in the selection of components. Each team of four students chooses three cities in their state and plans a trip incorporating each of the circumstances required in the plan. See *Rules for Tri-City Travel*. The design of the trip and its report requires lots of trial and error and problem solving with multiple operations of whole numbers and decimals.

Management
1. Some consideration needs to be directed toward the formation of working teams. They may be assigned by the teacher or perhaps include some direction from the teacher and some student choice too. For example, perhaps each student could choose one person with whom he or she would like to collaborate and the teacher completes the team by his/her design.
2. Since the task is open-ended, it will be important to describe appropriate and approved resources for students to access—travel agents, Internet access, magazines, newspapers, etc.

3. In order to keep the groups moving along on their projects, establish some benchmarks along the way. These benchmarks could be determined with the help of the students or the teacher could pre-determine them. They could include choice of three cities, list of first five resources to be contacted or gathered, and general description of proposed trip (for approval by teacher).
4. Each group will need three copies of the *Budget Ledger*. They may also need multiple copies of the journal page.
5. A set of icons has been included for optional use on the timetable page.

Procedure
1. Describe the task to students and help them form working teams of four.
2. Discuss the task with the students and respond to questions for clarification. For example, the kind of van students choose will affect the gas mileage.
3. Thoroughly discuss the nature of the problem and a variety of resources that are acceptable. Guidelines for Internet use should be established.
4. Generate a list of resources that could be helpful to students as they consider the problem and the choices they have. For instance, travel agents or the state automobile organization could help provide information about lodging, entertainment, and food. The yellow pages of a phone book or the travel section of a newspaper or magazine are also good sources of helpful information. All resources need to be approved by the teacher before use.
5. Inform the class of deadlines along the way for key pieces of the project.

* Reprinted with permission from *Principles and Standards for School Mathematics*, 2000 by the National Council of Teachers of Mathematics. All rights reserved.

Tri-City Problem Solving

Rules for Tri-City Travel

Task Description

Your team is to design and account for a trip for the four of you that includes three cities of your choosing in your state. Each team will have an allotment of $2,000 and some fixed costs. There are also some events that are required to be included.

Task Elements

Three cities (including home city) must be included in the trip.

1. One stop must include some form of live entertainment. (In other words, not a movie, video or arcade games)
2. One stop must include a visit to a site that contributes to some academic/artistic growth and appreciation such as a museum, a zoo, or a national park, or a historical site or landmark.
3. The trip must account for travel, food, and lodging including mileage, gas, and incidentals. A car (a mid-sized Van or SUV and driver/chaperone) is provided and the cost to each team is a standard $250.

Presentation of Trip Design

Upon completion of the design of the virtual trip, a complete report including cost analysis and pictures and detailed explanation will be shared to an interested audience of students and adults. Be creative!

Four components are part of the final report. Each team member assumes responsibility for the completion of one piece. All members must have input into all sections and must give final approval.

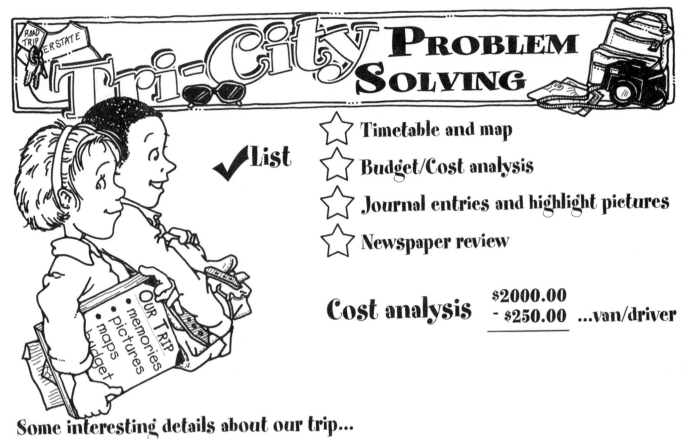

Tri-City PROBLEM SOLVING

✔ List
- ☆ Timetable and map
- ☆ Budget/Cost analysis
- ☆ Journal entries and highlight pictures
- ☆ Newspaper review

Cost analysis

$2000.00
- $250.00 ...van/driver
_____

Some interesting details about our trip...

Tri-City PROBLEM SOLVING

Date	Destination/Purpose	Arrival Time	Departure Time	Mileage/ Gas	Other Planned Stops

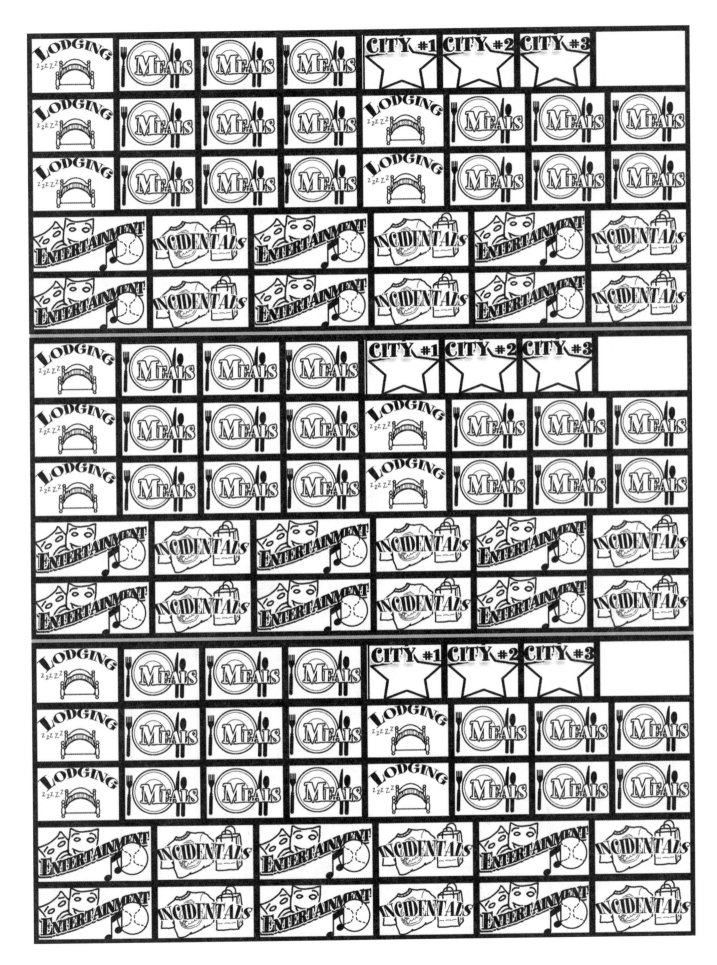

BUDGET LEDGER

2. Budget: List total costs for each item and calculate cost per individual. Provide a line item entry for each cost and its description under the appropriate headings. Other headings may be added at the discretion of the teacher or team.

City:	Line item entries/description	Total Cost	Individual
LODGING			
MEALS			
MILEAGE /GAS			
ENTERTAINMENT			
INCIDENTALS			

3. Journal entries and pictures for required highlights:

Describe in a journal or memory book the highlights of the planned stops along the way. Be sure to include pictures and descriptions that are creative and provide enough information to be appealing to the reader.

4. Newspaper review of historical/academic site:

Write a newspaper article describing the visit of the team to a historical site or museum. Include at least one picture and describe the significance of the visit to the team, the most interesting or surprising feature, and a general description of the site.

100 Multiplication Facts

5 x 8	7 x 3	3 x 9	5 x 4	0 x 4	3 x 8	4 x 1	5 x 0	9 x 9	4 x 7
6 x 8	2 x 7	1 x 4	9 x 3	5 x 3	2 x 2	5 x 1	0 x 9	2 x 8	1 x 7
3 x 2	1 x 2	1 x 0	8 x 5	7 x 2	5 x 9	7 x 9	3 x 7	8 x 4	5 x 5
2 x 3	3 x 1	8 x 2	0 x 1	8 x 1	1 x 5	2 x 0	1 x 6	2 x 5	8 x 3
6 x 0	1 x 8	6 x 6	4 x 8	9 x 6	7 x 8	2 x 4	3 x 3	7 x 0	6 x 2
4 x 0	8 x 9	4 x 9	7 x 5	0 x 5	9 x 8	0 x 8	4 x 2	1 x 9	4 x 5
4 x 4	2 x 0	9 x 1	0 x 2	7 x 6	6 x 5	8 x 6	6 x 9	1 x 1	5 x 2
2 x 6	7 x 1	0 x 6	2 x 1	3 x 0	9 x 5	5 x 6	9 x 7	9 x 4	7 x 4
9 x 2	6 x 1	8 x 0	3 x 5	1 x 5	3 x 4	7 x 7	6 x 4	0 x 0	3 x 6
0 x 7	4 x 6	6 x 3	5 x 7	9 x 0	4 x 3	8 x 7	6 x 7	0 x 3	8 x 8

90 Division Facts.

8)48	5)25	2)6	7)21	5)0	3)27	9)9	3)6	9)18
6)12	8)40	3)12	1)9	6)18	8)0	5)40	4)20	1)6
4)36	1)1	7)28	6)54	3)18	1)4	2)8	2)10	2)16
4)12	2)2	6)36	4)16	2)14	3)15	9)81	5)20	7)14
1)7	2)12	1)8	2)4	5)35	1)5	2)0	9)0	3)21
5)10	4)4	6)6	6)42	8)24	8)16	5)15	8)32	4)24
3)9	5)30	1)3	6)0	9)27	6)30	7)0	6)24	4)32
1)0	5)45	7)56	3)0	7)63	9)63	1)2	7)7	9)72
5)5	4)28	6)48	7)35	9)54	2)18	7)49	3)3	8)8
8)56	9)36	9)45	7)42	4)0	8)64	4)8	3)24	8)72

FINAL ANSWERS Multiplication I

Directions

Using the graph paper provided, draw the described arrays and find the area code or the answer to the multiplication problem.

1. 2×6 = _____

2. 4×7 = _____

3. 3×14 = _____

4. 5×16 = _____

Directions

Use the Multiplication Stretch way to solve these problems.

5. 65
   $\times 3$

6. 48
   $\times 5$

7. 67
   $\times 4$

8. 24
   $\times 6$

9. 157
   $\times 6$

10. 403
    $\times 3$

Directions

Solve these short stories. Use the back of the paper to draw pictures of your thinking.

Simple

1. 8 ft. by 12 ft. sleeping room
    How much carpet?

2. 3 gallons of mint chocolate chip ice cream
    16 scoops per gallon.
    How many single scoop cones?

FINAL ANSWERS

Division I

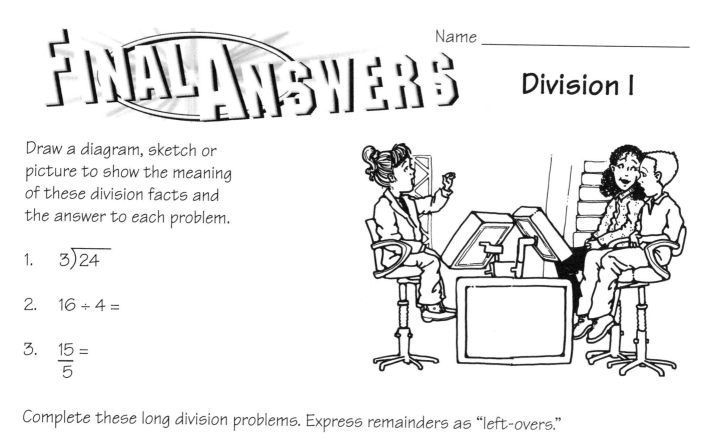

Name _____

Draw a diagram, sketch or picture to show the meaning of these division facts and the answer to each problem.

1. $3\overline{)24}$

2. $16 \div 4 =$

3. $\dfrac{15}{5} =$

Complete these long division problems. Express remainders as "left-overs."

4. $3\overline{)73}$ 5. $2\overline{)137}$ 6. $4\overline{)208}$ 7. $3\overline{)2601}$

Solve these short stories. Show your thinking processes on the back of the paper.

8. Fifty-one tennis balls on the court.
   How many "cans" to collect and store them? (Hint: 3 balls fit in each can.)

9. Soccer uniforms for the team. 3 yards of material per uniform. 127 yards of fabric on sale. How many outfits?

10. Eighteen American flags. How many stars? How many stripes?

FINAL ANSWERS

Draw a diagram, sketch or picture to show the meaning of and solution to these problems.

1. $8 \overline{)56}$

2. $49 \div 7 =$

3. $\dfrac{37}{4} =$

Solve these long division problems. Express remainders as fractions.

4. $7 \overline{)468}$ 5. $9 \overline{)6032}$ 6. $13 \overline{)4635}$ 7. $24 \overline{)3051}$ 8. $56 \overline{)25438}$

On the back of this page, solve these short stories. Show your thinking and problem-solving strategies.

9. 985 miles to go. 62 mph. How many hours? How many days?

10. Apple plant packed 6,724 apples. Twenty-eight to a carton. How many cartons?

The AIMS Program

AIMS is the acronym for "**A**ctivities **I**ntegrating **M**athematics and **S**cience." Such integration enriches learning and makes it meaningful and holistic. AIMS began as a project of Fresno Pacific University to integrate the study of mathematics and science in grades K-9, but has since expanded to include language arts, social studies, and other disciplines.

AIMS is a continuing program of the non-profit AIMS Education Foundation. It had its inception in a National Science Foundation funded program whose purpose was to explore the effectiveness of integrating mathematics and science. The project directors in cooperation with 80 elementary classroom teachers devoted two years to a thorough field-testing of the results and implications of integration.

The approach met with such positive results that the decision was made to launch a program to create instructional materials incorporating this concept. Despite the fact that thoughtful educators have long recommended an integrative approach, very little appropriate material was available in 1981 when the project began. A series of writing projects have ensued, and today the AIMS Education Foundation is committed to continue the creation of new integrated activities on a permanent basis.

The AIMS program is funded through the sale of books, products, and staff development workshops and through proceeds from the Foundation's endowment. All net income from program and products flows into a trust fund administered by the AIMS Education Foundation. Use of these funds is restricted to support of research, development, and publication of new materials. Writers donate all their rights to the Foundation to support its on-going program. No royalties are paid to the writers.

The rationale for integration lies in the fact that science, mathematics, language arts, social studies, etc., are integrally interwoven in the real world from which it follows that they should be similarly treated in the classroom where we are preparing students to live in that world. Teachers who use the AIMS program give enthusiastic endorsement to the effectiveness of this approach.

Science encompasses the art of questioning, investigating, hypothesizing, discovering, and communicating. Mathematics is the language that provides clarity, objectivity, and understanding. The language arts provide us powerful tools of communication. Many of the major contemporary societal issues stem from advancements in science and must be studied in the context of the social sciences. Therefore, it is timely that all of us take seriously a more holistic mode of educating our students. This goal motivates all who are associated with the AIMS Program. We invite you to join us in this effort.

Meaningful integration of knowledge is a major recommendation coming from the nation's professional science and mathematics associations. The American Association for the Advancement of Science in *Science for All Americans* strongly recommends the integration of mathematics, science, and technology. The National Council of Teachers of Mathematics places strong emphasis on applications of mathematics such as are found in science investigations. AIMS is fully aligned with these recommendations.

Extensive field testing of AIMS investigations confirms these beneficial results:

1. Mathematics becomes more meaningful, hence more useful, when it is applied to situations that interest students.

2. The extent to which science is studied and understood is increased, with a significant economy of time, when mathematics and science are integrated.

3. There is improved quality of learning and retention, supporting the thesis that learning which is meaningful and relevant is more effective.

4. Motivation and involvement are increased dramatically as students investigate real-world situations and participate actively in the process.

We invite you to become part of this classroom teacher movement by using an integrated approach to learning and sharing any suggestions you may have. The AIMS Program welcomes you!

AIMS Education Foundation Programs

Practical proven strategies to improve student achievement

When you host an AIMS workshop for elementary and middle school educators, you will know your teachers are receiving effective usable training they can apply in their classrooms immediately.

Designed for teachers—AIMS Workshops:
- Correlate to your state standards;
- Address key topic areas, including math content, science content, problem solving, and process skills;
- Teach you how to use AIMS' effective hands-on approach;
- Provide practice of activity-based teaching;
- Address classroom management issues, higher-order thinking skills, and materials;
- Give you AIMS resources; and
- Offer college (graduate-level) credits for many courses.

Aligned to district and administrator needs—AIMS workshops offer:
- Flexible scheduling and grade span options;
- Custom (one-, two-, or three-day) workshops to meet specific schedule, topic and grade-span needs;
- Pre-packaged one-day workshops on most major topics—only $3,900 for up to 30 participants (includes all materials and expenses);
- Prepackaged *week-long* workshops (four- or five-day formats) for in-depth math and science training—only $12,300 for up to 30 participants (includes all materials and expenses);
- Sustained staff development, by scheduling workshops throughout the school year and including follow-up and assessment;
- Eligibility for funding under the Eisenhower Act and No Child Left Behind; and
- Affordable professional development—save when you schedule consecutive-day workshops.

University Credit—Correspondence Courses

AIMS offers correspondence courses through a partnership with Fresno Pacific University.
- Convenient distance-learning courses—you study at your own pace and schedule. No computer or Internet access required!

The tuition for each three-semester unit graduate-level course is $264 plus a materials fee.

The AIMS Instructional Leadership Program

This is an AIMS staff-development program seeking to prepare facilitators for leadership roles in science/math education in their home districts or regions. Upon successful completion of the program, trained facilitators become members of the AIMS Instructional Leadership Network, qualified to conduct AIMS workshops, teach AIMS in-service courses for college credit, and serve as AIMS consultants. Intensive training is provided in mathematics, science, process and thinking skills, workshop management, and other relevant topics.

Introducing AIMS Science Core Curriculum

Developed in alignment with your state standards, AIMS' Science Core Curriculum gives students the opportunity to build content knowledge, thinking skills, and fundamental science processes.
- *Each* grade specific module has been developed to extend the AIMS approach to full-year science programs.
- *Each* standards-based module includes math, reading, hands-on investigations, and assessments.

Like all AIMS resources these core modules are able to serve students at all stages of readiness, making these a great value across the grades served in your school.

For current information regarding the programs described above, please complete the following:

Information Request

Please send current information on the items checked:

____ *Basic Information Packet* on AIMS materials   ____ Hosting information for AIMS workshops
____ *AIMS Instructional Leadership Program*   ____ AIMS Science Core Curriculum

Name _____   Phone _____

Address_____
     Street   City   State   Zip

AIMS Program Publications

Actions with Fractions, 4-9
Awesome Addition and Super Subtraction, 2-3
Bats Incredible! 2-4
Brick Layers II, 4-9
Chemistry Matters, 4-7
Counting on Coins, K-2
Cycles of Knowing and Growing, 1-3
Crazy about Cotton, 3-7
Critters, 2-5
Electrical Connections, 4-9
Exploring Environments, K-6
Fabulous Fractions, 3-6
Fall into Math and Science, K-1
Field Detectives, 3-6
Finding Your Bearings, 4-9
Floaters and Sinkers, 5-9
From Head to Toe, 5-9
Fun with Foods, 5-9
Glide into Winter with Math and Science, K-1
Gravity Rules! 5-12
Hardhatting in a Geo-World, 3-5
It's About Time, K-2
It Must Be A Bird, Pre-K-2
Jaw Breakers and Heart Thumpers, 3-5
Looking at Geometry, 6-9
Looking at Lines, 6-9
Machine Shop, 5-9
Magnificent Microworld Adventures, 5-9
Marvelous Multiplication and Dazzling Division, 4-5
Math + Science, A Solution, 5-9
Mostly Magnets, 2-8
Movie Math Mania, 6-9
Multiplication the Algebra Way, 6-8
Off the Wall Science, 3-9
Out of This World, 4-8
Paper Square Geometry:
    The Mathematics of Origami, 5-12
Puzzle Play, 4-8
Pieces and Patterns, 5-9
Popping With Power, 3-5
Positive vs. Negative, 6-9
Primarily Bears, K-6
Primarily Earth, K-3
Primarily Physics, K-3
Primarily Plants, K-3

Problem Solving: Just for the Fun of It! 4-9
Problem Solving: Just for the Fun of It! Book Two, 4-9
Proportional Reasoning, 6-9
Ray's Reflections, 4-8
Sensational Springtime, K-2
Sense-Able Science, K-1
Soap Films and Bubbles, 4-9
Solve It! K-1: Problem-Solving Strategies, K-1
Solve It! 2nd: Problem-Solving Strategies, 2
Solve It! 3rd: Problem-Solving Strategies, 3
Solve It! 4th: Problem-Solving Strategies, 4
Solve It! 5th: Problem-Solving Strategies, 5
Spatial Visualization, 4-9
Spills and Ripples, 5-12
Spring into Math and Science, K-1
The Amazing Circle, 4-9
The Budding Botanist, 3-6
The Sky's the Limit, 5-9
Through the Eyes of the Explorers, 5-9
Under Construction, K-2
Water Precious Water, 2-6
Weather Sense: Temperature, Air Pressure, and Wind, 4-5
Weather Sense: Moisture, 4-5
Winter Wonders, K-2

Spanish Supplements*
Fall Into Math and Science, K-1
Glide Into Winter with Math and Science, K-1
Mostly Magnets, 2-8
Pieces and Patterns, 5-9
Primarily Bears, K-6
Primarily Physics, K-3
Sense-Able Science, K-1
Spring Into Math and Science, K-1

* Spanish supplements are only available as downloads from the AIMS website. The supplements contain only the student pages in Spanish; you will need the English version of the book for the teacher's text.

Spanish Edition
Constructores II: Ingeniería Creativa Con Construcciones
    LEGO® 4-9
    The entire book is written in Spanish. English pages not included.

Other Publications
Historical Connections in Mathematics, Vol. I, 5-9
Historical Connections in Mathematics, Vol. II, 5-9
Historical Connections in Mathematics, Vol. III, 5-9
Mathematicians are People, Too
Mathematicians are People, Too, Vol. II
What's Next, Volume 1, 4-12
What's Next, Volume 2, 4-12
What's Next, Volume 3, 4-12

For further information write to:
AIMS Education Foundation • P.O. Box 8120 • Fresno, California 93747-8120
www.aimsedu.org • 559.255.6396 (fax) • 888.733.2467 (toll free)

Duplication Rights

Standard Duplication Rights

Purchasers of AIMS activities (individually or in books and magazines) may make up to 200 copies of any portion of the purchased activities, provided these copies will be used for educational purposes and only at one school site.

Workshop or conference presenters may make one copy of a purchased activity for each participant, with a limit of five activities per workshop or conference session.

Standard duplication rights apply to activities received at workshops, free sample activities provided by AIMS, and activities received by conference participants.

All copies must bear the AIMS Education Foundation copyright information.

Unlimited Duplication Rights

To ensure compliance with copyright regulations, AIMS users may upgrade from standard to unlimited duplication rights. Such rights permit unlimited duplication of purchased activities (including revisions) for use at a given school site.

Activities received at workshops are eligible for upgrade from standard to unlimited duplication rights.

Free sample activities and activities received as a conference participant are not eligible for upgrade from standard to unlimited duplication rights.

Upgrade Fees

The fees for upgrading from standard to unlimited duplication rights are:
* $5 per activity per site,
* $25 per book per site, and
* $10 per magazine issue per site.

The cost of upgrading is shown in the following examples:
* activity: 5 activities x 5 sites x $5 = $125
* book: 10 books x 5 sites x $25 = $1250
* magazine issue: 1 issue x 5 sites x $10 = $50

Purchasing Unlimited Duplication Rights

To purchase unlimited duplication rights, please provide us the following:
1. The name of the individual responsible for coordinating the purchase of duplication rights.
2. The title of each book, activity, and magazine issue to be covered.
3. The number of school sites and name of each site for which rights are being purchased.
4. Payment (check, purchase order, credit card)

Requested duplication rights are automatically authorized with payment. The individual responsible for coordinating the purchase of duplication rights will be sent a certificate verifying the purchase.

Internet Use

Permission to make AIMS activities available on the Internet is determined on a case-by-case basis.

* P. O. Box 8120, Fresno, CA 93747-8120 *
* aimsed@aimsedu.org * www.aimsedu.org *
* 559.255.6396 (fax) * 888.733.2467 (toll free) *